LEARNING AND TECHNOLOGICAL CHANGE

Also by Ross Thomson

THE PATH TO MECHANIZED SHOE PRODUCTION IN THE UNITED STATES

Learning and Technological Change

Edited by

Ross Thomson
Associate Professor of Economics
University of Vermont, Burlington

St. Martin's Press

First published in Great Britain 1993 by
THE MACMILLAN PRESS LTD
Houndmills, Basingstoke, Hampshire RG21 2XS
and London
Companies and representatives
throughout the world

A catalogue record for this book is available
from the British Library.

ISBN 0–333–55683–6

Printed in Great Britain by
Antony Rowe Ltd
Chippenham, Wiltshire

First published in the United States of America 1993 by
Scholarly and Reference Division,
ST. MARTIN'S PRESS, INC.,
175 Fifth Avenue,
New York, N.Y. 10010

ISBN 0–312–09591–0

Library of Congress Cataloging-in-Publication Data
Learning and technological change / edited by Ross Thomson.
p. cm.
Includes index.
ISBN 0–312–09591–0
1. Technological innovations—Economic aspects. 2. Technological innovations—Economic aspects—History. 3. Economic history.
I. Thomson, Ross.
HC79.T4L423 1993
338'.064—dc20 92–46516
CIP

Contents

List of Tables and Figures

Tables

Figures

Acknowledgements

This book brings together the research of fifteen scholars who interpret technological change as an institutionally structured learning process. Like the change it examines, this book was the outcome of a learning process. It began as a working conference on 'The Process of Technological Change' held at the Graduate Faculty of the New School for Social Research in November 1989. In this lively and cooperative setting, the authors interacted with each other, with other presenters (including Paul David, Gerard Dumenil, Thomas Hughes, Peter Lazes, Dominique Levy, William Mass, Scott Moss, Pascal Petit, Nina Shapiro and Edward Wolff) and with participating faculty and students. Like industrial research, the conference needed financial support, for which we thank the New School. It also needed organizational support, which Karin Ray and a dozen volunteering students provided with care and enthusiasm.

Just as a new technique requires development after the inventor shouts 'Eureka', so too did the essays of this book. Authors modified their papers, often substantially. John Berry read and insightfully commented on the entire manuscript. Edward Nell and Floria Thomson helped to clarify the introduction and epilogue. The editors at Macmillan gave the book a consistent style

As the book argues, learning does not stop when a technique is commercialized. Nor does it stop when a book is published. Users and readers alike learn from the product, and their feedback advances the knowledge of inventors and authors. That a successful technique leads to further new techniques is a theme of this book. To advance knowledge in a way that brings about further advances is the book's goal.

ROSS THOMSON

Notes on the Contributors

Alice H. Amsden is Leo Model Professor of Economics and Professor of Political Science at the Graduate Faculty, New School for Social Research. Her latest book is *Asia's Next Giant: South Korea and Late Industrialization*.

Alfred D. Chandler Jr is Straus Professor of Business History, Emeritus, Harvard Business School. He has taught at the Massachusetts Institute of Technology and Johns Hopkins University. His major works are *Strategy and Structure: Chapters in the History of the American Industrial Enterprise, The Visible Hand: The Managerial Revolution in American Business*, and *Scale and Scope: The Dynamics of Industrial Capitalism*. He was Assistant Editor of *The Letters of Theodore Roosevelt*, 4 vols (1952–54), and Editor of the *Papers of Dwight D. Eisenhower*, 5 vols (1970).

Francesca Chiaromonte is a doctoral student at the University of Minnesota.

Carolyn C. Cooper, Research Affiliate at Yale University, is a historian of technology. She is the author of *Shaping Invention* and of many articles, in such journals as *Technology and Culture* and *IA*, on manufacturing technology and patenting in the nineteenth century.

Giovanni Dosi is Professor of Economics at the University of Rome 'La Sapienza'. Previously he was Research Fellow, Science Policy Research Unit, University of Sussex. He is the author of several works on the economics of technical change and evolutionary economics, including *Technical Change and Industrial Transformation*, and was coeditor of *Technical Change and Economic Theory*.

Donald Harris has been Professor of Economics at Stanford University since 1972. His major publications include *Capital Accumulation and Income Distribution*, and numerous articles in leading economics journals on capital theory, growth theory, income distribution and economic development.

Takashi Hikino is Research Associate of Business History at the Graduate School of Business Administration at Harvard University. He has published in the fields of business history and economic development.

William Lazonick is Professor of Economics at Barnard College, Columbia University, and Research Associate at the Harvard Institute for International Development. He was formerly on the faculties of the Harvard Economics Department and Harvard Business School. His recent books include *Competitive Advantage on the Shop Floor*, *Business Organization* *and the Myth of the Market Economy* and *Organization and Technology in Capitalist Development*.

Edward J. Nell is Professor of Economics at the New School for Social Research in New York. He has taught at Weslyan University and the University of East Anglia, and has been Visiting Professor at the Universities of Bremen, Paris, Rome, Siena, Orleans and Nice. A Fellow of the Lehrman Institute, a Bard Center Fellow, and Fellow and teaching member of the Centro di Studi Economici Avanzati, he is on the editorial board of the *Journal of Post Keynesian Economics*. He is the author of *Prosperity and Public Spending*, *Transformational Growth and Effective Demand* and *Rational Economic Man* (with Martin Hollis) and the editor of *Growth, Profits and Property* and *Free Market Conservatism*.

Richard R. Nelson is Professor of International and Public Affairs, Business and Law at Columbia University. He is the author, with Sidney G. Winter, of *An Evolutionary Theory of Economic Change* and the editor of *National Innovation Systems: A Comparative Study*.

Luigi Orsenigo is Associate Professor of Economic Policy at the Bocconi University, Milan, Italy. After completing his post-graduate studies at the Science Policy Research Unit, University of Sussex, he has been teaching and working at the University of Brescia and at the Bocconi University. His publications include *The Emergence of Biotechnology*.

William N. Parker is Phillip Golden Bartlett Professor, Emeritus, of Economics and Economic History at Yale University. He was one of the last students of Abbott Payson Usher, the notable technological historian, at Harvard. He taught American and European economic history for 30 years at the University of North Carolina and at Yale. A collection of his writings has been published in two volumes as *Europe, America, and the Wider World*.

Willi Semmler is Associate Professor of Economics at the New School for Social Research, New York. He has taught at the Free University of Berlin and the American University, Washington, DC, and has been Visiting Professor at the Université d'Orleans, the Universities of Bremen and Mexico (UNAM) and Visiting Scholar at Columbia University in New York, CEPREMAP in Paris, and Stanford University in Palo Alto, California. He is the author of *Competition, Monopoly, and Differential Profit Rates*, the editor of *Competition, Instability, and Nonlinear Cycles, Financial Dynamics and Business Cycles*, and coeditor, jointly with Edward Nell, of *Nicholas Kaldor and Mainstream Economics*.

John K. Smith is Associate Professor of History at Lehigh University. He is the author, with David A. Hounshell, of *Science and Corporate Strategy: Du Pont R&D, 1902–1980*.

Ross Thomson is currently Associate Professor of Economics and the chairperson of the Department of Economics at the University of Vermont. From 1976 to 1991, he was Assistant Professor of Economics and then Senior Lecturer in Economics at the New School for Social Research. His publications include *The Path to Mechanized Shoe Production in the United States* and numerous articles examining the process of technological change and its relation to the historical evolution of capitalist economies.

1 Introduction

Ross Thomson

Technological change has become a major public concern. Some countries face problems of declining competitiveness, stagnant wages and regular trade deficits. Many more face pressures from slowing growth of productivity and national income. Third World countries see technological change as a means to develop, while former Eastern bloc countries explore the relation of the market to productivity growth. Governments everywhere try to formulate productivity-enhancing policies. Clearly, in order to understand economic trends and formulate constructive policy we must understand technological change.

Economists and other social scientists have a parallel concern. As their once-static focus has taken a dynamic turn, economists have found that technological change is a principal source of economic growth and rising per capitum income. Students of business identify it as basic cause of the growth of the corporation. Its effects on employment, the distribution of income and regional differences in growth are carefully scrutinized. The slowdown and, for some countries, relative decline of the past two decades have focused attention on the limits to technical change, increasingly replacing the celebratory tone of writing during the long post-war boom by a tone of anxious and perplexed urgency.

However, this confluence of public interest and academic expertise has not yet led to adequate explanations of technical change. The reasons are many. Too often tied to a static theory in which technology is, as Nathan Rosenberg puts it, a black box, economists have charted consequences of technical change without similar penetration into its sources. Technical change has proved too complex to enable the formulation of compelling models from the many useful case studies written by social scientists. To this must be added the modest communication among the economists, economic historians, sociologists, business historians and historians of technology who have best addressed the problem of technological change.

The contributions in this book make inroads into this problem by interpreting technological change as a learning process. Learning is an issue that crosses the spectrum of the social sciences, as these studies, coming from microeconomics, macroeconomics, economic history, business history and the history of technology, illustrate. Although varying in method and object, each chapter argues that individuals in institutional settings – including firms, industries, patenting systems and occupations – act in ways that lead them to learn. Learning fosters invention that alters production. More or less complex feedback mechanisms thus lead from current techniques and organizations of production to new kinds.

The promise of this interpretation is that it does not have to rely on outside changes to bring about new techniques. Rather, we can look to learning processes that emerge from the functioning of basic institutions. Technological change may form virtuous circles in which conducive institutions generate learning and innovations that strengthen the institutions. Through this cumulative character, technical change may build on itself in much the way Adam Smith and his followers argue that market growth may extend the division of labour, which increases productivity, which in turn expands the market. In this interpretation, institutional differences can account for differences in productivity growth among countries and time periods. For this reason the essays in this book, although not aimed directly at the public concerns of the present, carry important implications for addressing these concerns.

Several demanding requirements must be met to fulfil this promise. Learning arguments must identify circumstances conducive to such learning. Economic history reveals long periods when experience in production did not lead to productivity growth. What, then, are the kinds of social organization that support learning and innovation, and what kinds stifle learning or prevent its translation into new techniques? Much has been written about whether markets or planning better supports technical change, but we must also distinguish the roles of kinds of markets (such as capital goods, intellectual property or labour) and kinds of plans (such as central, intrafirm or interfirm).

Learning arguments must also explain why some techniques change but not others. Most efforts to change techniques concern relatively minor improvements, rather than more radical changes. An adequate notion of technical change would account for each kind and their relationship. This would help us understand the growing diversification of firms in the twentieth century. It might also help explain economy-wide phenomena such as the historical leadership of the USA in mass-production technologies.

Finally, an adequate explanation must show how technical change can continue over time and across industries and localities. Once industrialization began, capitalist economies showed a remarkable capacity to generate a stream of technological changes. At the same time, any particular technology may have limited scope for improvement, and firms and whole economies have become locked into constraining technologies. Learning may help us understand the continuity of technical change by explaining how and with what limits particular technologies evolve.

The essays of this book advance several connected theses to meet these requirements. First, learning and its results depend on the institutional context of those who learn. This context can support or block the interactions through which individuals learn and translate learning into new techniques. As this book will show, firms that differ in their organization of production, integration with marketing and research, and purposes for inventing also vary in their involvement in technical change. These differences imparted distinct features to British, American, German and Japanese economies, and thus are of more than scholarly concern. Industry structures, occupations, patent systems and other institutions all have similar significance.

Second, successful learning and innovation involve flows of knowledge among institutions. Innovating firms hire workers given technological training in other firms and in educational institutions. Flows of information from potential and actual users of new techniques provide direction for inventive activity. Just as innovators learn from people and institutions across society, they also – often quite unintentionally – teach across an equally wide gamut. Firms, governments, professions and educational establishments affect the extent to which technological knowledge is shared by structuring communications channels.

Third, learning has a private dimension that allows firms to benefit from their invention. In developing new techniques, innovators depend on knowledge, organization and practices specific to their institution. Their sales and servicing staffs learn from users of the new technique, but this learning does not easily flow to their competitors. Competitors find that duplicating the innovation entails costs of reinvention. Successful technological change thus combines private and public dimensions; countries that lead technologically have secured both the substantial growth of innovating firms and the rapid spread of technological knowledge.

The fourth thesis maintains that the standard, linear sequence from invention to development to commercialization must give way to a feedback view in which sales, licensing and use generate ongoing learning and invention. The purposes of inventors change in the course of invention. Firms often modify products in response to consumer opinion; the spread of nascent techniques can thus influence their technological development. Radical inventions often succeed when new kinds of consumers are discovered; radios were only wireless telephones before commercial broadcasting. Competition from other innovators can hamper, stimulate or redirect invention.

A fifth thesis is that changes in technique are often bound up with changes in institutions. As several essays show, technical change may accelerate when firms are reorganized. Furthermore, institutions commonly stimulate those kinds of technical change that reinforce these institutions, much as the British long remained leaders in some small-batch engineering. Institutional innovation may be necessary to develop other kinds of technological change, but established institutions, well used to once-successful techniques, may find these innovations particularly difficult. As US businesses are currently discovering, institutional innovations can give advantages to latecomers, penalize earlier organizational forms and transform competition from national to global scales.

These theses are interwoven throughout the four parts of this book. The first part identifies three complementary approaches to technological change. The study of technological change is too often separated by academic discipline and method, and as a result fails to make use of common insights. The essays of this part, written from the perspectives of the economist, the business historian and the historian of technology, identify distinct institutions and processes that are all needed to understand the process of technological change.

The second part examines the connections among diffusion of techniques, learning and ongoing technological change. By too sharply dividing the generation of new techniques from their spread, scholars and policy-makers misunderstand how economic processes can generate new techniques. Developing a practical technique depends on learning from inventors elsewhere in the economy and from users of nascent techniques. Diffusion is also a costly learning effort and one that, in the context of appropriate institutions, fosters further technological change.

The third part analyses the competitive aspects of technological change. Firms, industries and countries generate and diffuse techniques at different rates, and this has profound competitive consequences as our current concern with changing competitiveness of nations makes evident. But the connection goes deeper; competitive success depends on the capacity to learn, and competitive relations influence this capacity and whether firms invest in developing it.

The last part focuses on interrelated histories of technical change and institutional evolution. Institutional restructuring has often altered learning processes; changes in the pace of technological change, firm structure and economic growth can result. The international character of this history is more and more critical and, in a more integrated world, competitive advantages by some countries require quicker responses by others.

This volume has a history that may help clarify the idea of learning. The volume's theme is itself an outcome of a learning process. In 1988, the economics department of the New School for Social Research began to organize a conference on 'The Process of Technological Change'. The conference had two premises: to explain the pace and direction of technical change, we have to understand the process generating and diffusing new techniques; and to best advance this understanding, we should bring together scholars who approach the question from a variety of usually isolated perspectives.

Two dozen scholars met in November 1989. Their previous work analysed technological change from a wide range of perspectives: economic history, industrial organization, macroeconomics, microeconomics, economic development, sociology, management, business history, the history of technology, and labour. The discussion made clear that many participants had formulated arguments about the learning process which often offered considerable support for each other. In a summarizing session, participants agreed that a book interpreting technological change as a learning process would bring together most of the conference debate and would contribute to a broader understanding of technological change. The contributions of this book, often greatly revised from the originals, are the outcome.

Of course, intellectual and technological advances are quite different in character. Whereas one involves scholars, universities, disciplines and publication, the other concerns inventors, firms, technologies and occupations.

But both involve institutions, communication, feedback and complementary organizational changes. Moreover, just as inventors hope that their efforts will alter techniques of production, so the authors of this book hope that their research will alter the way we understand the transformation of production techniques.

Part I
Approaches and Institutions

Economists, business historians, and historians of technology approach technological change differently. Yet as the essays of this part demonstrate, each of these approaches adds important elements to an interpretation of technological change as a learning process.

Economists have noted that technological change can be inhibited by the uncertainty that investing in new knowledge will bring returns and by the appropriation of these returns by free-riding imitators. Richard R. Nelson addresses these issues in his interpretation of technological change as 'cultural evolution'. He argues that technological knowledge has specific and generic aspects. In developing successful techniques, firms form tacit knowledge, which cannot be codified easily or spread quickly. This firm-specific knowledge provides advantages that justify firms' investment in knowledge production. Because of firm-specific knowledge and uncertainty, firms develop a variety of new techniques. Competition then selects the 'fittest' technique. The generic aspects of technological knowledge spread readily through the activities of firms and technological communities, such as craftsmen and engineers. Knowledge-sharing institutions, which give technical change a cultural dimension, accelerate the pace and determine the kinds and geographic scope of technological change.

The principal institution bringing new techniques to fruition is of course the firm. Alfred D. Chandler Jr relates the innovation process to the organization of the firm in a study of US, British and German industrial companies. Established firms were involved less with initial invention than with development and commercialization. Such firms were most successful when they designed and organized effective systems to produce and market the new technique or its products. Not only did these complementary innovations in firm structure help to spread the new technique, but they also communicated possible improvements to the firm. Profits and knowledge accumulated in spreading the technique in turn fostered further technological change. Successful firms institutionalized innovation in industrial research laboratories that supported the cycle of diversification, firm growth and ongoing technical improvements.

Firms innovate in an environment of other institutions, and at times those institutions play a key role in technical change. Carolyn C. Cooper focuses on institutions of patent management, which secure, enforce, renew and gain use for patents. Using the methods of the history of technology, she argues that in the pre-Civil War USA these institutions comprised 'an invisible college of technology'. They spread knowledge of new techniques across industries and locations. Moreover, they educated prospective inventors, entrepreneurs and the general population about how to patent and also, by shaping ideas about the identification and description of technological novelty, how to invent. Such education was particularly important prior to organized industrial research.

2 Technical Change as Cultural Evolution

Richard R. Nelson

This chapter develops the following themes. First, technical advance needs to be understood as a cultural evolutionary process. Models by economists that treat it as a maximizing process miss the point, as do the models that treat it as just like a model in biology. Second, what provides private incentive to invest in research and development (R&D) is that the fruits are at least partly a private good. However, new technology has a public aspect as well, and this is what makes the evolutionary process cultural. As a result, technology advances through the work of a community of technologists who are both rivals and colleagues. Third, to a good first approximation the private part of technology is the specific application, and the public part generic. This is so for two reasons: the means that firms have to protect their inventions largely shield the specifics, but the generic aspects of their inventions are almost impossible to protect; and in addition, universities are scientific and technical societies and are in the business of making generic knowledge public. Fourth, technological communities (and the technological change they engender) have become more science based, and more transnational, in recent times. The consequences are profound.

TECHNICAL ADVANCE AS A CULTURAL EVOLUTIONARY PROCESS

My basic theoretical starting point is the proposition that technical change must be understood as a cultural evolutionary process (Nelson and Winter, 1982). It is an evolutionary process in that inevitably it proceeds through the generation of a variety of new departures in competition with each other and with prevailing practice. There are real winners and losers in that competition, and these are determined not through *ex ante* calculation but in actual contest.

While economists who have studied the details of technical advance understand this well, and it is also the standard view of the process in other disciplines (for example, history and sociology) that have studied it, most contemporary formal models of technical advance in economics miss the point. Thus virtually all the models of technical innovation recently reviewed by Reinganum (1989) treat technical change as a process that can be predicted and planned in advance. They ignore the fundamental uncertainty that surrounds any effort to make a significant move forward in technology, and the differences in judgement as to what will and will not succeed.

Different inventors and companies lay their bets differently. Some succeed and some do not.

It would be a mistake, however, to view the evolutionary process by which technical advance proceeds as just like evolutionary processes in biology. While there are a number of differences, the most important one is that new findings, and new useful ways of doing things, do not adhere strictly to their finder or inventor and well-defined descendants but are shared among contemporaries, at least to some extent. The capabilities of all are advanced by the creation or discovery of one. The process is one of cultural evolution, which is fundamentally different from evolution in the biological sense.

The best known theoretical statement of this point of view of course is Schumpeter's: 'the essential point to grasp . . . is that we are dealing with an evolutionary process' (Schumpeter, 1950, p. 82). Schumpeter clearly recognized the 'cultural' aspects of that process. While in his model the lure and reward for private innovative efforts reside in a temporary monopoly over the new product or process, he stressed that (in the general run of things) monopoly is temporary. Sooner or later a company's proprietary technology will be found out and imitated or invented around. He did not see this as a problem, but rather as a mechanism to spread the benefits of successful innovation widely and to keep companies from resting on their laurels.

A number of modern economists, recognizing this matter, have tried to analyse it in terms of a trade-off between incentives for innovation, which are seen as hinging on the ability of firms to make new technology proprietary, and the extent to which an innovation is employed where it is efficient to use it, which may depend on the degree to which new technology is public. The trade-off language is used explicitly in the writings on optimal patent length.[1] But it is implicit in much of the writing on 'spillovers' from one firm's R&D to competitors (see, for example, Nelson, 1988). In this literature, the inability of a firm to keep its competitors from free-riding dampens R&D incentives while at the same time extending the range of use of the new technology. Undoubtedly there is a lot to this point of view, but it misses several important things.

First, the fact that new technology goes public not only increases society's ability to use it in its present form, but also widens the range of parties who are in a position to improve it, variegate it and more generally contribute to its further advance. Some analysts (such as Kitch, 1977) have put forth the argument that technical advance may proceed with less waste in a setting where one firm can control the path of development. However, while there surely are wastes, technical advance generally has proceeded much more rapidly in settings where a number of competitors were involved than where one or a few parties controlled development.[2]

In such a context it is true up to a point at least that spillovers can enhance incentives for R&D rather than diminish them. In the model of Levin and Reiss (1989) and that of Cohen and Levinthal (1989), spillovers and own

R&D are complements (see also Romer, 1990). Thus in an industry as a whole the direct diminution of incentives for R&D because of incomplete appropriability can be partially offset by the stimulus to R&D stemming from ability to build on the results gained by others.

Second, simple analyses of this genre do not recognize adequately that technology is not a unitary thing but can be unpacked. The most important issue may be what *aspects* of an invention (or technology, more generally) are private, and what *aspects* are public. It has been argued elsewhere (Nelson, 1982) that technology at any time needs to be understood as consisting of both a set of specific designs and practices, and a body of generic knowledge that surrounds these and provides understanding of how things work, key variables affecting performance, the nature of currently binding constraints and promising approaches to pushing these back. Dosi (1982) has used the term 'technological paradigm' for these broad bodies of technological understanding which illuminate practice.

A particular aircraft design, the Boeing 767, draws on the broad body of generic knowledge aircraft designers have in their heads and libraries. In turn, because there is a community of aircraft designers that follows each other's work, knowledge of the particular design characteristics and performance of the 767 adds to the general understanding used by designers as they go about their work. Companies design and produce specific semiconductor chips, but much of the knowledge and techniques that go into a particular chip design is generic in the sense of bearing on a wide class of chip designs. In turn a successful new chip design provides information that expands knowledge more broadly. In the same sense, one can distinguish between a particular drug used in chemotherapy, and the knowledge about drugs generally and about chemotherapy possessed by doctors, scientists and hospital technicians. To a considerable extent these different aspects of technology can be treated differently in law and custom.

It is tempting to think of generic technological knowledge as 'science' and the body of specific practice as 'applications'. However, while in some cases generic technological knowledge is science-like, it is a mistake to think of the set of operative designs and techniques as applied science. In many cases where it is known that a technique works, it is not well understood how. Thus medical scientists do not understand why many effective drugs (for example, aspirin) have the effects they do. The effectiveness of a particular technique, as with the therapeutic properties of aspirin, is often discovered initially through trial and error, or even accident. It may take a long time before any generic understanding is developed of why and how it works. Indeed, wrong theories may persist for some time. For example, Lee de Forest clearly had mistaken ideas about what was going on in his triode.

Further, in most fields a considerable portion of generic understanding stems from operating and design experience with products and machines and their components, and generalizations reflecting on these. Thus consider a mechanic's guide, or the general knowledge of potters, or steelmakers, or the shared lore of aircraft designers. Much of generic technological knowledge has only limited grounding in any fundamental science, standing as it were largely on its own foundations. This is not quite what most lay persons mean when they talk of a 'science'.

However, a number of observers have noted that many modern fields of inquiry that call themselves sciences do not fit the conventional image. Thus fields like thermodynamics, metallurgy, pathology and the various engineering disciplines are basically oriented towards practical understanding, and reflect attempts to make such knowledge more scientific.[3] These 'applied sciences' are dedicated to developing an understanding of technology that goes deeper than practitioners' knowledge, and often they reach down into fundamental physics, chemistry and biology to provide a basis. However, by no means have they been able to encompass and codify all generic knowledge, much of which in many fields continues to be largely in the experience of engineers and technicians. And to a considerable extent the applied sciences have analytic structures of their own which are not completely reducible to the underlying fundamental sciences.

In much of economic theory it is presumed that technology is a latent public good, in the sense of having the capacity to benefit many parties, and being inexpensive (if not literally costless) to teach and learn compared with the cost of invention or discovery in the first place. Now this notion is a reasonable first approximation if the focus is on generic knowledge. Generic knowledge not only tends to be germane to a wide variety of uses and users, but technological communities and professions tend to grow up around such knowledge, which then becomes the stock in trade of the insiders of a field. There then tends to grow up a systematic way of describing and communicating such knowledge, so that when new generic knowledge is created anywhere it is relatively costless to communicate to other professionals.

On the other hand, the set of particular known techniques is more of a mixed bag. Some of practised technique is widely applicable and easily learned by someone skilled in the art, if access is open. But scholars like Keith Pavitt (1987) and Nathan Rosenberg (1976, 1982) have argued persuasively that many prevailing industrial techniques which operate effectively in a given establishment can be transferred to another only with considerable cost, even if the original operator is open and helpful. Partly that is because of the complexities involved and the high cost of teaching and learning the details. Partly it is because significant modifications may need to be made if the technology is to be effective in a somewhat different context. In many cases specific practice is buried deep in the structure and custom of the host institution and is difficult to disentangle. In these cases technology, or at least a particular technique, is more like a private good than a latent public good.

This is the first part of the argument about what aspects of technological advance are private and what aspects public. To a good first approximation, new specific applications are private, at least when they are new. On the other hand, when technological communities are relatively well defined and institutionalized, new generic technological knowledge, including the general principles associated with new proprietary products and processes, is not only a latent public good but a manifest one. New generic understanding spreads rapidly among the knowledgeable community, and can be used by all.

There certainly are exceptions. Generic knowledge of potentially wide significance won through the R&D efforts of a company can be secret for a long time under certain circumstances, and in some cases patents have been defined and enforced so broadly as to, in effect, give the patent holder control over the use of a significant area of generic technology. The infamous Seldon automobile patent is a good example, and some modern biotechnology patents, while not yet tested in court, seem to give the holder such broad gauged control (see Mergis and Nelson, 1990). My caveat, 'at least when they are new', signals that I recognize that many specific techniques rapidly become public to those skilled in the art. But while there are important exceptions, I propose (as a reasonable first approximation and a useful basis for economic analysis and modelling) that the distinction between generic knowledge and specific applications corresponds to the public–private split regarding technology.

There are two reasons for the correspondence. First, the means firms have to appropriate returns from their technical innovations largely protect the particular applications. Indeed, in many instances little or no active 'protection' of a particular product design or method of production is necessary because direct imitation is innately very difficult. On the other hand, the advance in generic knowledge won through a firm's R&D is very hard to protect. Second, the professions defined by the applied sciences, and the university system that teaches new scientists and engineers and also engages in considerable research in the applied fields, provides an expressly public context for generic technological knowledge. On the other hand, university faculties tend to have a patchy knowledge of the details of specific industrial practice. The following two sections are concerned with supporting these claims. The concluding section draws some implications.

THE APPROPRIATION AND EXPROPRIATION OF PROPRIETARY QUASI RENTS

How in fact do firms appropriate returns from product and process innovation? Different models assume different mechanisms. Several years ago I and my colleagues, Richard Levin, Alvin Klevorick and Sidney Winter, devised a questionnaire designed to explore the question empirically and systematically. The details of the study and a survey of some of the key findings have been reported elsewhere (especially in Levin *et al.*, 1987), and further reports are forthcoming. Here my purpose simply is to show how those findings shed light on the question of *what* is private about technology.

To oversimplify somewhat, we distinguished four broad classes of means through which firms can appropriate returns to their innovations – through the patent system, through secrecy, through exploiting various advantages associated with lead time, and through supporting their innovation with complementary investments such as sales and service efforts – and asked our

Table 2.1 The effectiveness of various means for capturing returns to innovation: number of industries giving high scores

	Product innovation		*Process innovation*	
	5 or higher	*6 or higher*	*5 or higher*	*6 or higher*
Patents for protection	36	9	12	1
Patents for royalties	23	7	9	5
Secrecy	9	7	33	11
Lead time	103	47	85	28
Learning curve lead	82	26	78	20
Sales and service efforts	115	57	53	19

N = 127.

respondents in different lines of business to score on a scale from one to seven the effectiveness of these for protecting product and process innovations. The responses regarding product innovations will be our focus here. The responses regarding process innovations are similar in many respects.

There were significant cross-industry differences regarding the means rated most effective for appropriating returns to product innovation. However, as shown in Table 2.1, in most industries lead time (and such associated potential advantages as ability to move down the learning curve), and establishing an effective sales and service effort, were rated as the most effective means. Included here were such industries as semiconductors, computers, telecommunications and aircraft. In some of these industries patents were also rated as reasonably effective, in others not. But clearly in many industries where innovation is important, the gains to an innovator come largely from getting in early and supporting that advantage through complementary investments. Teece (1986) argues and empirically supports the same position.

There were several industries in which patents were rated as quite effective. In some of these, principally those producing complex chemical products, it would appear that without patent protection R&D would not pay.[4] However, there are not many industries where patents were the basic instrument for appropriating returns. And while there are exceptions, patents for the most part have been construed as protecting particular products and processes as contrasted with broad areas of generic technology. (For an analysis of patent scope see Mergis and Nelson, 1990, and Scotchmer, 1991). In no R&D intensive industry was secrecy rated an effective means to protect product innovation.

Table 2.2 Imitation cost as a percentage of innovation cost for new products: number of industry responses

	Less than 50	*51–75*	*76–100*	*More than 100*
Major patented new product	5	17	63	42
Major unpatented new product	18	58	40	11
Typical patented new product	20	64	32	11
Typical unpatented new product	67	40	15	5

$N = 127$.

The reasons firms find it worthwhile to invest in significant amounts of R&D, despite the fact that neither patents nor secrecy is generally effective, is that (contrary to views in some parts of economic analysis) in most industries the specific technology of a new product is not a public good, even a latent one. Table 2.2 shows the distribution of reports by industry on the cost of imitating relative to innovating both when the product is patented and when it is not, for major innovations and for more routine ones. Imitation costs tend to be high, even if the innovation is not patented.

However, this by no means indicates that firms are ignorant about their rivals' product innovations and technologies. We asked specifically about how firms acquire knowledge about product innovations developed by their competitors. The means included doing independent R&D or reverse engineering, licensing the technology, learning through patent disclosures and publications of various sorts, attending open technical meetings, conversations with employees of the innovating firms and, perhaps, hiring them away. Table 2.3 reports the responses.

The responses show (as argued by Cohen and Levinthal, 1989, among others) that monitoring outside technological developments generally is an active and costly business. In most industries the means of monitoring product innovations judged most effective was either doing independent R&D (presumably while attending to clues about what one's competitors are doing) or reverse engineering. The industries that gave these means low scores almost invariably were those that do little R&D themselves and hence which do not have the capabilities to employ them. Conversely, virtually all R&D-intensive industries rated one or both very effective as a means of learning about (and presumably mastering something comparable to) one's competitor's innovations. It is apparent that in these industries the fact that viable firms have active R&D efforts serves to bind them together technologically, as well as to advance the frontiers.

Just what do firms learn about their competitors' new product and process technology through their active monitoring efforts? Discussions with corporate R&D executives indicate that the objective seldom is to enable a firm to mimic directly, which may be blocked by patents or simply be impossible;

Table 2.3 Effectiveness of different means of learning about a competitor's product innovation: number of industries giving high scores

	5 or higher	*6 or higher*
Licensing	17	4
Patent disclosures	24	5
Publications or open mechanical meetings	20	8
Consultations with employees of the innovating firm	21	8
Hiring away innovator's employees	33	8
Reverse engineering	65	22
Independent R&D	84	19

rather, it is to learn enough so that the firm can apply the lessons to its own design and development efforts. Thus the result is not a clone but something somewhat, perhaps significantly, different. (For an analytic characterization, see Lippman and Rumelt, 1982.)

These facts about the means firms use to appropriate returns to innovation and to monitor each other's innovations puts firms' policies towards secrecy in a particular light. They suggest that secrecy is tight mainly *before* a new product gets introduced, so as to establish a lead time or patent position, but that it is likely to be looser afterwards. This is confirmed by the data in Table 2.3 showing that many of our respondents report they learn a lot from conversations with employees of innovating firms, and from publications and meetings of technical societies. It appears that, contrary to common beliefs, firms do not keep tight controls on all information about their new technology, and in some cases they seem actively to divulge information (see also von Hippel, 1988). Why is this?

In the first place, the very staking of claims involves the release of information. That is one of the intents of the patent system and, where patents are effective, they also make knowledge public. Where patents are not effective, but aggressive use of a head-start advantage is, firms have strong incentives to stake their claims through advertising, open meetings and a wide variety of other ways. They need to attract customers, and this means they need to tell them about their new wares, but this also means leaking something to their competitors.

Further, the divulging of generic technological information generally does not significantly undermine a company's real proprietary edge. Where new product patents are effective, as in pharmaceuticals, it does not hurt a company to let out generic information if it has the patent. It is noteworthy that scientists in pharmaceutical companies, for whom patents are very important in protecting new products, publish extensively in scientific journals (see Koenig, 1983). Even where patents are not strong, letting generic knowledge won in R&D go free does not prevent a firm reaping handsomely from its product innovation if it has a significant head-start on production and marketing of the product in question and the capacity to take advantage of that lead.

Let me summarize. The particular product and process designs that come out of a firm's R&D effort tend to be hard for competitors to imitate cheaply and quickly. In a few industries the reason is that patents are effective, but in most it is that these are innately quasi-private goods. On the other hand, the general undertandings behind those particular products and processes are difficult to keep privy for very long in an industry where most firms have sophisticated and alert engineers and scientists, and where these tend to communicate with each other. Advances in generic understanding won through corporate R&D efforts thus go public quite quickly, but this does not mean that rival firms get a free ride. They must do their own R&D in order to create their own specific ways of implementing this new knowledge.[5]

THE APPLIED SCIENCES AND THE UNIVERSITIES

The existence of a body of generic knowledge about a technology known to a community of practioners does not presume the existence of any underlying formal science. For as long as humans have made things and practised a trade there has been lore relating to the practice which newcomers had to learn before they could become effective. Reasonably well-defined craft communities long ago grew up to structure the passing on of such knowledge to new entrants and to regulate entry. The medieval guilds were defined around such bodies of generic knowledge as well as technique, and were designed to keep outsiders from that knowledge and at the same time to educate new members of a trade. In many technologies today generic knowledge continues to be in the heads and fingers of those who work with the technology, rather than in textbooks and articles, and hence learned on the job rather than in school.

The strength of modern science, and especially the rise of the applied sciences and engineering disciplines as fields of academic teaching and research, has profoundly changed the way generic technological knowledge is taught, learned and advanced.

In the USA and other countries university science and engineering and the science-based industries grew up together. Chemistry took hold as an academic field at about the same time as chemists began to play an important role in industry. The rise of university research and teaching in the field of electricity paralleled the creation of an electrical equipment industry. The new engineering disciplines were specially structured to provide industry with people with appropriate training (see, for example, Noble, 1977, and Landau and Rosenberg, 1989). An academic research tradition began to grow up in these fields, consciously aimed at facilitating the advance of technology.[6]

Particularly since 1945, research at universities has played a vital role in technological advance. The electronic computer and modern biotechnology

Table 2.4 Number of industries citing university research in a field as relevant to industrial technology

Field of science	*5 or more*	*6 or more*
Biology	12	3
Chemistry	19	3
Geology	0	0
Maths	5	1
Physics	4	2
Applied maths and Operation and Research	16	2
Computer science	34	10
Metallurgy	21	6
Materials science	29	8
Agricultural science	17	7
Medical science	7	3
Chemical engineering	19	6
Electrical engineering	22	2
Mechanical engineering	28	9

were born in universities.[7] While these cases are exceptional, university research now is a strong help to technical advance in a large number of industries. There has grown up a myth that the contributions of academic research to technical advance are seldom intended by the academics. In a few cases, like the early advances in modern biotechnology, the practical applications were seen only dimly *ex ante*. In the case of the computer, however, both the researchers and their funders had their eyes very clearly on the creation of practical technology. This is much more typical. Today, where academic research contributes relatively directly to technical advance, this largely comes about because the academics are deliberately trying to facilitate technical advance.

In our questionnaire, respondents were asked to score the relevance of university research in various fields to technical advance in this line of business on a scale of one to seven. Our fields included the basic sciences of physics, mathematics and chemistry as well as the applied sciences and engineering disciplines. Table 2.4 shows the number of industries that gave academic research in a field a relevance score of five or higher. The applied sciences – computer science, metallurgy and material science (a modern relation of metallurgy), and the engineering disciplines – head the list. Academic research in chemistry, a more conventional science, was also scored as relevant by a number of industries, which is not surprising to anyone familiar with the long-standing close links between academic and industrial chemistry.

Note that few of our respondents said academic research in physics or mathematics was important to technical advance. One must understand these responses as reflecting a certain myopia, since historians of science and technology know better. However, the responses do suggest that the impact of academic research in the basic science fields occurs over the long run, and is indirect, through influences on the applied sciences and engineering disciplines, which are the fields that are directly hooked into facilitating technical advance.

In the applications-oriented sciences and engineering disciplines, research is to large extent motivated to facilitate technical advance. Many academics are tuned into prevailing best-practice technology. Knowledge of its strengths and its weaknesses helps them to define their own research agendas. Indeed, in fields such as electrical engineering and biochemistry, the lines between what academics do and what some corporate scientists and engineers do may be blurred, not sharp. The vast bulk of corporate R&D is dedicated to the design, development and improvement of particular new products and processes. Academics do virtually none of this. However, a certain portion of corporate R&D, which varies from industry to industry and firm to firm, is more broadly oriented. Corporate scientists may be charged with gaining understanding relevant to solving certain practical problems, or exploring in areas where it is judged that a breakthrough would open up significant new technological opportunities.[8] Here there is a great deal of overlap between what the corporate scientists and engineers do, and what the academics do.

Thus the academics in the applied sciences and industrialists trained in a field and doing research in it together form a symbiotic professional community. They attend professional meetings together. They trade and share the scientific and technological news. The presence of these disciplines and professions is a strong force in making generic technological knowledge public.

THE CHANGING NATURE OF TECHNOLOGICAL COMMUNITIES

The fact that specific applications of a technology are largely private, while generic knowledge is mostly public among professionals in the field, helps one to see and to understand certain phenomena that otherwise tend to be ignored or puzzling. In particular, it enables one to understand why technology, like science, tends to advance with many actors contributing and building on each other's work, despite the fact that technologists, unlike scientists, tend to try to make the fruits of their work proprietary. Thus one can read virtually any history of a significant technology as a story about competition and about cooperation, because in fact the history involves both. This is just what I meant when I stated that technical change was a cultural evolutionary process.

Technological fields clearly differ in terms of the closeness of efforts to advance practice to ongoing science. Some years ago Derek Price (1965) argued that, for the most part, the science needed to advance technology was the 'packed down' science of the textbooks and handbooks, and mostly what the industrial scientist or engineer had learned in school, rather than frontier science. However, Price observed that technologies relating to chemistry and electronics were exceptions, with technical advance often drawing on (or being intertwined with) frontier sciences. The project at Bell Labs that led to the transistor, and a Nobel prize for several of those involved, is a signal example. Recent work by Narin and Noma (1985) and Narin and Frame (1989) has attempted to quantify the links between science and technology by examining the extent to which patents in a field cite the recent scientific literature. Their work confirms the judgement by Price about the relative closeness to frontier science of the chemical and electronic industries.

Electronics and the chemical technologies undoubtedly remain exceptional in their intertwining with frontier science. However, there is good reason to believe that, as a result of sustained support of the applied sciences since Price's time, many technologies now are closer to science than they used to be. Thus Narin and Frame (1989) show that over the last decade in a wide variety of fields there has been a striking increase in the number of science articles cited in patents, and the average 'age' of the cited article has shrunk.

This development strongly suggests that professionally trained scientists and engineers, who stay up with their fields, have become increasingly important in technological advance. At the same time, long hands-on experience in firms has become less important as a privileged vehicle to gain generic understanding. As a result generic technological knowledge has become more public to the professional scientific and technical community, and less a monopoly of those working in firms with long experience with the technology. The economic implications are profound.

In Thomson's wonderful study of the evolution of shoe machinery in the USA (1989), the community generally consisted of American mechanics who had learned their skills on the job. Similarly, to a considerable extent the development of Bessemer steel-making technology in Great Britain and in the USA proceeded through the work of British and Americans who knew about making steel because they were involved in the process, rather than through any school learning. They formed invention communities because what each had accomplished was embodied in artefacts available for scrutiny by others, not because they read the same journals. The communities that formed through familiarity with artefacts and processes tended to be geographically proximate, and often natural.

On the other hand, if one looks at the history of dyestuffs or radio, the picture is quite different. For one thing, citizens of a number of different countries were involved, and built on each other's work. For another, the 'inventors' generally were university trained in the relevant underlying disciplines. These two aspects are closely related. The presence of a strong underlying science, generally an applied science or engineering discipline, makes generic technological knowledge world knowledge, rather than national knowledge. When the underlying generic knowledge is strongly codified, taught in universities and written up in scientific journals, a technologist in one part of the world may be able to make a pretty good guess about just how a particular invention made in another part of the world works, even though the reports he or she has of it may be very sketchy.

It is apparent that more and more technological histories are coming to look like the history of dyestuff chemistry and of radio, and fewer like that of shoemaking machinery and nineteenth-century steel-making. The relevant technology communities now are university-trained professionals rather than persons who have learned their technology on the job, and are global rather than national. With the globalization of technology, the economic development of nations now is tied together as never before.

Notes

Support for the research that led up to this chapter has been provided by the National Science Foundation (NSF), and by the Sloan Foundation under its grant to the Consortium on Competition and Cooperation. David Mowery, Nathan Rosenberg and Eric von Hippel have provided useful feedback on earlier versions. Alvin Klevorick, Richard Levin and Sidney Winter were my partners in designing and executing the questionnaire that provided the data used here to support the thesis I put forth. And, as in so many matters, my ideas in this chapter are intertwined with those about evolutionary theory more generally that I share with Sidney Winter.

1. Nordhaus (1969) provides the most complete statement.
2. See Mergis and Nelson (1990) for an examination of several industrial histories with this question in mind. Cohen and Levin (1989) review the literature on the relationships between technical advance and market structure.
3. Rosenberg has stressed the importance of these 'non-classical' sciences, particularly in Chapter 7, 'How Exogenous is Science', in his *Inside the Black Box* (1982). See also his more recent 'Why Do Companies Do Basic Research With Their Own Money' (*Research Policy*, 1990), and Mowery and Rosenberg (1989). See Noble (1977) and Landau and Rosenberg (1989) on the rise of the engineering disciplines.
4. Our findings regarding where patents are important are similar to those reported earlier in Scherer *et al.* (1959), Mansfield *et al.* (1971) and Wyatt, Berlin and Pavitt (1985).
5. For an excellent discussion of the problems firms have when they rely on external sources of R&D and do not invest in their own, see Mowery and Rosenberg (1989), Ch. 5.
6. Noble (1977) and Thackray (1982) provide useful accounts.
7. On computers see Flamm (1988). Kenney (1986) is good on the biotechnology story.
8. For a similar argument see Rosenberg (1990) and Pavitt (1987).

References

Cohen, W. and Levin, R. (1989) 'Empirical Studies of Innovation and Market Structure', in R. Schmalensee and R. Willig (eds), *Handbook of Industrial Organization* (Amsterdam: North-Holland).

—— and Levinthal, D. (1989) 'Innovation and Learning: The Two Faces of R and D', *The Economic Journal* (September).

Dosi, G. (1982) 'Technological Paradigms and Technological Trajectories', *Research Policy* (Summer).

Flamm, K. (1988) *Creating the Computer* (Washington, DC: Brookings Institution).

Kenney, M. (1986) *Biotechnology: The University Industrial Complex* (New Haven, Conn.: Yale University Press).

Kitch, E. (1977) 'The Nature and Function of the Patent System', *Journal of Law and Economics* (October).

Koenig, E. (1983) 'A Bibliometric Analysis of Pharmaceutical Research', *Research Policy*.

Landau, R. and Rosenberg, N. (1989) 'Successful Commercialization in the Chemical Process Industries', Stanford University, Mimeo.

Levin, R. and Reiss, P. (1988) 'Cost-reducing and Demand-increasing R&D with Spillovers', *Rand Journal of Economics* (Winter).

Levin, R., Klevorick, A., Nelson, R. and Winter, S. (1987) 'Appropriating the Returns to Industrial R and D', Brookings Papers on Economic Activity, 1987–3.

Lippman, S. and Rumelt, R. (1982) 'Uncertain Imitability: An Analysis of Interfirm Differences in Efficiency Under Competition', *Bell Journal of Economics* (Autumn).

Mansfield, E., Rapoport, J., Schnee, J., Wagner, S. and Hamburger, M. (1971) *Research and Development in the Modern Corporation* (New York: Norton).

Mergis, R. and Nelson, R. (1990) 'The Complex Economics of Patent Scope', *Columbia Law Review* (May).

Mowery, D. and Rosenberg, N. (1989) *Technology and the Prospect of Economic Growth* (Cambridge University Press).

Narin, F. and Frame, J. (1989) 'The Growth of Japanese Science and Technology', *Science* (11 August).

——— and Noma, F. (1985) 'Is Technology Becoming Science?, *Scienceometrics*.

Nelson, R. (1982), 'The Role of Knowledge in R&D Efficiency', *Quarterly Journal of Economics* (August).

Nelson, R. (1988) 'Modelling the Connections in the Cross Section Between Technical Advance and R and D Intensity', *Rand Journal of Economics* (Autumn).

—— and Winter, S. (1982) *An Evolutionary Theory of Economic Change* (Cambridge, Mass.: Belknap Press).

Noble, D. (1977) *America by Design* (New York: Knopf).

Nordhaus, W. (1969) *Innovation, Growth, and Welfare* (Cambridge, MA: MIT Press).

Pavitt, K. (1987) 'On the Nature of Technology', Inaugural lecture given at the University of Sussex, 23 June.

Price, D. de S. (1965) 'Is Technology Historically Independent of Science? A Study in Statistical Historiography', *Technology and Culture*, 6 (Autumn).

Reinganum, J. (1989) 'The Timing of Innovation: Research, Development, and Diffusion', in R. Schmalensee, and R. Willig (eds), *Handbook of Industrial Innovation* (Amsterdam: North-Holland).

Romer, P. (1990) 'Endogenous Technical Change', *Journal of Political Economy*, 98.

Rosenberg, N. (1976) *Perspectives on Technology* (Cambridge University Press).

—— (1982) *Inside the Black Box: Technology and Economics* (Cambridge University Press).

—— (1990) 'Why do Companies do Basic Research with their own Money?' *Research Policy*.

Scherer, F. (1959) *Patents and the Corporation* (Boston, privately printed).
Schumpeter, J.A. (1950) *Capitalism, Socialism, and Democracy* (New York: Harper & Row).
Scotchmer, S. (1991) 'Standing on the Shoulders of Giants', *Journal of Economic Perspectives* (Winter).
Teece, D. (1986) 'Profiting From Technological Innovation', *Research Policy* (December).
Thackray, A. (1982) 'University-Industry Connections and Chemical Research: An Historical Perspective', *University–Industry Research Relationships* (Washington, DC: National Science Board).
Thomson, R. (1989) *The Path to Mechanized Shoe Production in the United States* (Chapel Hill, NC: University of North Carolina Press).
Von Hippel, E. (1988) *The Sources of Innovation* (Oxford University Press).
Wyatt, S., with Bertin, G. and Pavitt, K. (1985) 'Patents and Multinational Corporations: Results from Questionnaires' *World Patent Information* (Autumn).

3 Learning and Technological Change: The Perspective from Business History

Alfred D. Chandler Jr

At the core of business history is the firm. There the practices and activities of business are developed, modified and reshaped. There the processes of industrial production, distribution, communication and finance are carried on in modern capitalistic economies. And there much of the learning so central to technological change takes place. To paraphrase Joseph Schumpeter, capitalism has been an engine for technological change and for-profit firms in rivalrous competition are the featured actors in this evolutionary drama (1942, Chapters 6–8). The *primary* role of the business firm in technological change has been that of the *development*, not the discovery, of new products and processes.

As Richard Nelson emphasized (see Chapter 2), in the post-Second World War years the university and the government have also become significant players in technological change. Universities and institutes and, more so in the past than in the present, individual scientists, engineers and inventors have carried out research and invention. But the universities and individual inventors rarely transformed research into widely used products and, if technological innovations are little used, they remain not much more than interesting intellectual feats.

Government agencies have done research, but over a smaller spectrum and with a narrower focus than universities. In fact, more innovative research has been carried by firms than by government agencies. National laboratories such as Livermore and Los Alamos have concentrated almost wholly on national defence projects. Since 1945 government agencies have been involved in developing new products and processes, but their contributions have been limited to a few specialized areas such as nuclear power, supersonic aircraft and space exploration.

The primary role of the federal government in R&D has been two-fold: as a funder of the research and as a purchaser of new products and processes. Since the Second World War such agencies as the NSF, the National Institutes of Health, the Department of Defense and the Department of Energy (and its predecessor the Atomic Energy Commission) have funded university research. As customers, the Defense and Energy Departments have provided funds for research and even more for the development of new products and processes. But the actual development, like the production of these technological innovations, has been carried out within firms.

The role of the firm in the development of new products and processes is particularly relevant for the themes of this book, as development – the bringing of technological innovations into widespread use – is a learning process. The transformation of an invention into an innovation requires continuous learning in defining product design, in determining and then scaling up the most cost effective processes of production, in analysing the markets and devising ways to reach them, in obtaining reliable sources of supplies and materials and in acquiring an efficient workforce and competent managers.

From the perspective of the business firm three different sorts of development can be discerned. First is the learning involved in the initial commercialization of a *new* technological innovation. Here learning often takes place as the firm itself is being created to carry out the commercialization of the innovation. Second is the institutionalizing of learning processes to develop new goods and machines through the formation of a separate specialized organization within an already established firm distinct from its other functional departments for production, distribution and obtaining supplies. Third is the continuous cumulative improvement of *existing* products and processes that occurs within an established firm in its functional departments. Because more attention has been given to the third of these three development activities in later chapters of this book, this chapter will concentrate on the first two (see Chapters 5, 6, 10 and 13; also Rosenberg, 1979, pp. 25–50). These two emphasize dramatically the close connection between the business firm, the learning process and technological change. Let us focus, then, first on the initial learning processes in the development of a new or greatly improved product or technology – a process that was often part of the creation of a modern integrated industrial enterprise – and then on the work of specialized units in which the learning process was systematized within the institution.

INITIAL LEARNING AND ITS INSTITUTIONALIZATION

The building of modern transportation and communications networks in Europe and the USA brought on a wave of technological innovations that transformed processes of production and distribution. The new methods of manufacturing and marketing, in turn, transformed the ways learning occurred in commercializing new products and processes.

The completion and integration of rail, telegraph, steamship and cable systems – a feat carried out by the very first of the new large managerial business enterprises – made possible an unprecedented volume, regularity and speed of flow of goods and messages through the economy. As a result, the output of manufacturing establishments increased sharply. For the first time producers were able to exploit economies of scale. That is, for the first time large plants were able to manufacture at lower per unit costs than smaller plants (with the minimum efficient size of plants being determined by the nature of the technology and the size of the existing markets). In addition unit costs decreased when several products were produced in the same set of facilities using much the same raw and semi-finished materials so that producers were able to exploit the economies of scope (Chandler, 1990a, Ch. 2, especially pp. 34–6).

In this context, the commercialization of new products involved more than merely setting up a manufacturing establishment to produce the new goods for sale. It required the scaling-up (and tight integration) of the different processes within the works if producers were to exploit effectively the cost advantages of economies of scale and scope. Such scaling-up and integration involved experimentation, false starts, and the shaping and reshaping of equipment before minimum efficient size of the high volume works was achieved. The process took time: at least eight years at the Lynn, Massachusetts, works of the Thomson Houston Electric Company (forerunner of General Electric: see Carlson, 1989, pp. 18–22); over four years at Henry Ford's Highland Park Plant to produce the Model T (Nevins, 1954, Ch. 18, pp. 504–8); over 10 years at Bayer's massive dye and pharmaceutical works on the Rhine at Leverkausen (Chandler, 1990a, pp. 477–8); five years at Du Pont at a much smaller dye-making works in Deepwater, New Jersey (Hounshell and Smith, 1988, pp. 87–97); and as long at Brunner Mond's (forerunner of Imperial Chemical Industries' synthetic ammonia plant at Billingham near Durham, England: see Reader, 1970, pp. 363–7).

At the same time, in order to market the resulting unprecedented volume of output, the entrepreneurs who formed the manufacturing firm had to create a national (and often international) sales organization. A sales force had to be recruited to locate and learn the needs of potential customers. This internalizing of marketing in turn had a powerful impact on product design and production methods. The feedback permitted what Ross Thomson has described and aptly termed 'learning by selling' (see Chapter 6).

With the integration of production and selling, the business firm became a conduit for continuing interaction between the technological processes and customer needs. Within the firm relationships were regularized between its marketers, its production managers, its plant and process designers, and often also between these managers and outside specialists in technology at engineering schools such as MIT and the California Institute of Technology, and at the engineering and scientific departments of state universities in the USA and the Technische Hochschulen in Germany.

At the centre of the initial learning process, therefore, were the full-time salaried managers who were responsible for scaling up and then operating the plants, who built and managed the marketing and distribution networks and, most important of all, the senior executives who recruited lower-level managers, coordinated and monitored their work, and arranged for the financing of the costly venture.

The most successful enterprises in commercializing the new products of what historians have properly termed 'the Second Industrial Revolution' were, then, those that made three sets of investments: in manufacturing plants large enough to exploit the cost advantages of scale and scope; in marketing and distribution large enough to reach national and international markets; and in a managerial hierarchy large and competent enough to direct and coordinate the enterprise's manufacturing and marketing activities and to plan and set policies for its present health and continuing growth.

Here it is important to distinguish between the *inventors* of a new product or process, the *pioneers* who initially commercialized the product or process, and the *first movers* who make investments in facilities and personnel large enough to utilize fully the economies of scale and scope. The first firms to make such three-pronged investments in these new capital-intensive, volume-producing industries acquired powerful competitive advantages. To benefit from comparable costs, challengers had to build plants of comparable size, as or after the first movers worked out the bugs in the new production processes. Challengers had to create distribution and sales organizations to capture markets where first movers had already established themselves. They had to recruit management teams to compete with those already well down the learning curve in their specialized activities in production, distribution and R&D. Challengers did appear, but they were few; and they were rarely successful until they had made a three-tiered investment in manufacturing, marketing and management comparable to those of the first movers. Not surprisingly, as Richard Nelson points out in Chapter 2, in most industries a first-mover strategy (that is, exploiting 'lead time [and such associated potential advantages as ability to move down the learning curve], and establishing an effective sales and service effort') was rated as 'the most effective means' for appropriating returns to product innovation, much more effective than either patents or secrecy.

From the 1880s, these first-movers and challengers were Schumpeter's for-profit firms which in rivalrous competition were the featured actors in the drama of technological change in the new capital-intensive industries. Once the first movers had made their investments, the structure of the industries they dominated became oligopolistic. A few large firms competed for market share and profits, less in terms of price than of functional or strategic capabilities. Price remained a valued competitive weapon, but more significant for competitive success were the abilities of firms to improve their processes of production and distribution, to build more effective labour relations, to provide more effective marketing services, to find more suitable suppliers, and to move more quickly into expanding markets and out of older declining ones. The test of the existence of such competition was changing market share; and in all these new oligopolistic capital-intensive industries market share and profits changed constantly.

In the more science-based, technologically complex industries – those enjoying the greatest cost advantages of scale and scope – many of the new enterprises soon institutionalized the learning acquired in the creation of the enterprise by building specialized organizations separate from the departments of production, sales, purchasing or finance. In such organizations the emphasis continued to be on development, not research. It was development that often required several years of constant effort and continuing learning before profits were realized. It was development that accounted for 80–90 per cent of the total R&D cost of getting a new product to market. Such institutionalization involved the development of routines for building prototypes, pilot plants and semi-works in production; for sampling and testing markets to understand specific needs and wants; for locating suppliers; for recruiting personnel with relevant skills (usually from within the firm, although sometimes from without); and finally for arranging the financing of the projects (primarily from the firm's own revenues.)

In seeking out new products and processes for which the firms' existing production and marketing skills might be used to bring in international markets, their development organizations (especially those of American and German firms) spread their nets widely. In addition to examining the ideas and projects that came from within the enterprise, their managers talked in early years to inventors, and later to more specialized research laboratories and institutes. They monitored the activities of small pioneering firms, including start-up companies which might be acquired and then scaled up. Their libraries and reference offices kept close tabs on relevant patents and articles in industry and scientific journals. Their managers were leaders – and in many cases founders – of professional and trade associations. They kept in close touch with the professional schools and universities from which many had graduated and on whom they relied for a wide range of technological and scientific information. In this way, as members of the growing technological community, they converted what Richard Nelson in Chapter 2 calls public or generic knowledge into private, proprietary knowledge. In this way the competition between large functionally integrated business firms which had institutionalized the learning process involved in the development of new goods and methods became a driving force for technological change in the capitalist economies of the West.

LEARNING FROM ENTERPRISE CREATION: HISTORICAL EXAMPLES

The significance for technological change of learning acquired in the creation of large-scale modern industrial enterprises is dramatically illustrated by the performance of British and German firms in the new industries of the Second Industrial Revolution. The British pioneers too often failed to make investments large enough to exploit the economies of scale and scope. Their failure not only deprived them of the institutional arrangements essential to participate in continuing technological change through the development of new products and processes, but even their ability to maintain existing technical competence.

The difference in the performance of the British and German chemical industries is particularly striking (Chandler, 1990a, pp. 278–9, 474–86). The earliest of the new science-oriented industries was the dye industry. The first man-made dye was invented and then commercialized by a Briton, William Perkin. Dyes were made from coal, and Britain had the largest supply of high-quality coal in Europe. Its huge domestic textile industry continued to be the world's largest market for the new dyes until at least the Second World War. Given such comparative advantages, British entrepreneurs should have soon dominated the world in this new industry. However, British entrepreneurs failed to build works large enough to exploit fully first the economies of scale and then those of scope. They continued to rely on outsiders to distribute and sell their products, and they recruited only a small number of salaried managers. They failed to learn how best to profit from the new ways of production and distribution.

In the late 1870s and 1880s three large German companies (Bayer, BASF and Hoechst) and three smaller firms made such investments. They built giant plants on the Rhine which, by utilizing economies of scale, brought the price of a kilo of dye down from well over 100 marks in the early 1870s to 23 marks in 1878. Then, by exploiting the economies of scope through the production of literally hundreds of different dyes in the same plants using the same raw and semi-finished materials, they brought the price down to 9 marks a kilo by 1886. At the same time they created worldwide marketing organizations, for every user of natural dyes had to be taught and often also had to be provided with special machinery to use the new products effectively. Thus before the end of the century Bayer had a global sales force of experienced chemists to maintain close contact with over 20 000 customers scattered throughout the world. By then its development laboratories were the best organic chemical laboratories in the world. By that time, too, the managerial hierarchies operating these firms were among the largest in the world.

The resulting competitive advantages of the German firms demolished Britain's comparative advantages. In 1913, 160 000 tons of dyes were produced. German companies made 140 000 (with 'the big three' accounting for 72 per cent of that output). Ten thousand more came from Swiss neighbours up the Rhine. The total British production was 4400 tons (Haber, 1971, pp. 145–79). The comparative performance of German and British enterprises was much the same in pharmaceuticals, films, agricultural chemicals and electro-chemicals as it was in organic chemicals. Therefore, while the Germans continued to bring on-stream a number of major new chemical products, the British, except for rayon, produced almost none until the 1930s, when the formation of Imperial Chemical Industries created an institution with the necessary capabilities for product development.

The history of the electrical equipment industry parallels the chemical story. British innovators were as active in this industry as any in Germany. Sir William Mather, the senior partner of Mather & Platt (one of the largest British textile manufacturers), obtained the Edison patents at the same time that Emil Rathenau did at AEG. Then in the 1880s and 1890s AEG and Siemens – and General Electric (GE) and Westinghouse in the USA – made the essential first-mover investments, but Mather & Platt did not.

In 1903, after merging with smaller competitors, both Siemens and AEG embarked on long-term plans to assure their position in world markets. Siemens systematized and rationalized production by concentrating its operations under a single management. It then built the world's largest industrial complex, a set of huge works covering several square miles that Siemens financed largely from retained earnings. (The municipality of Berlin that the complex dominated soon became officially known as Siemensstadt.) Where Bayer had built a single works, Siemens constructed several in which more than 20 000 workers made telecommunications equipment and instruments, dynamos and other heavy electric machinery, small motors, electro-chemicals and cables. In the same years AEG followed suit, constructing similar but somewhat less massive works at nearby Hennigsdorf. Well before 1903 both Siemens and AEG had worldwide sales forces of trained electrical and mechanical engineers. They had built their extensive laboratories for the development of new product lines and the improvement of old ones. Moreover, their managerial organizations were as impressive as those of the chemical industry and of their counterparts in the USA.

By 1913 two-thirds of the electrical equipment machinery made in British factories by British labour came from the subsidiaries of GE, Westinghouse and Siemens. AEG sold more products in Britain than the largest British company did, while Mather & Platt had become a minor producer of electrical equipment for factories. From the 1890s onwards, therefore, R&D to improve existing products and develop new ones was carried out in Schenectady, Pittsburgh and Berlin, but not in Britain (Chandler, 1990a, pp. 276–7, 464–73).

In metal-making the failure of British entrepreneurs to make the investments and build the organizations necessary to exploit the economies of scale meant that their firms in time lost the knowledge and technical skills essential to maintain technical competence. In copper the British failure to build in the 1890s plants large enough to exploit the economies of scale permitted by the new electrolytic refining process meant that, from that time on, Britain was no longer a major producer of finished copper. Because of the small investments British entrepreneurs made in aluminium works, their industry remained a secondary factor in international markets.

In steel, where British innovators (including Sir Henry Bessemer, Sidney Gilchrist and his cousin Sidney Thomas) led the way, the story is more complex. Given the size and nature of the market for steel there were rational reasons for not scaling up production facilities to the size of those in the USA and Germany. However, because such plants were not built, British firms lost their markets abroad, and then even at home. As a result, when concentrated efforts were made after 1918 by companies, banks and the government to rebuild the British steel industry, the leading firms – Stewarts & Lloyds, Colvilles, Richard Thomas and United Steel Companies – all went to the Chicago firm of H.A. Brassert & Co. for plans to build best-practice steel plants. The experience and the learned skills required to construct a modern steel works were no longer available in Britain (Chandler, 1990a, pp. 279–84, 320–32).

THE LEARNING PROCESS INSTITUTIONALIZED: HISTORICAL EXAMPLES

After the new enterprises had learned the intricacies of the new ways of production and distribution, many quickly institutionalized that product-specific learning. By systematizing the learning acquired in bringing their initial product on-stream, these firms were the first to create specialized internal organizations for product and process development. In Germany these activities were usually carried on in specialized 'laboratories', and in the USA normally in departments organized separately from the already-formed production, marketing and supply departments. Once such learning had been institutionalized, these large industrial enterprises began to develop – to commercialize – other products, often for new markets that used the specialized competencies learned in creating the enterprise (that is, they began to diversify into related lines).

Thus, the German dye firms quickly challenged German first movers in pharmaceuticals, including Merck and Schering (enterprises that are still powerful today). As early as the 1890s Bayer was bringing to the market sedatives and pain-killers, of which aspirin became the best known. Hoechst was even more innovative, developing Novocaine, fever-depressing drugs, the first serums for diphtheria, vaccines for cholera and tetanus and the first chemotherapeutical drug, Salvarsan, the earliest effective cure for syphilis. By the First World War these German chemical firms, particularly BASF with its Haber–Bosch process, used their development skills to produce new high pressure technology methods in the making of nitrates and other fertilizers at very low cost (Chandler, 1990a, pp. 478, 485).

War, defeat, occupation and inflation held back continuing new product development in German chemicals until after 1925. Then the time was brief before the Depression and Hitler's command economy again reduced funds and personnel available for development of non-military products. So during the 1920s and 1930s, American, rather than German, chemical and pharmaceutical enterprises became the generators of technological change in their industries. For example, Du Pont's venture into polymer chemicals (one of the few examples of fundamental research by an industrial enterprise before 1939) led to the development of a variety of man-made inexpensive and durable yarn and cloth (nylon, dacron, orlon and lycra). The output of this new chemical industry quickly overtook that of the world's largest and oldest industries, those based on the processing of natural fibres: cotton, wool, silk and linen. In the 1970s synthetic production matched natural fibre production, and by 1980 more than twice as much synthetic fibre was consumed as the natural fibre products. In the interwar and early post-war years Du Pont, through its development organizations, commercialized moisture-proof cellophane, refrigerants, automotive finishes, synthetic rubber, film, plastics and other materials, and became a major player in agricultural chemicals, electrochemicals, paints, plastics and other products (Hounshell and Smith, 1988, p. 384, also Chs 13, 18 and 19). John Smith's contribution (see Chapter 5 below) indicates the complex scientific and technological relationships involved in such development as well as the market/technology interaction.

Other American chemical firms followed suit, diversifying by commercializing products based on their core technologies: Dow on salt chemistry (one new project was, for example, volume-produced magnesium), Monsanto on saccharine, Union Carbide on carbon, Koppers on coke, American Cyanamid on calcium and Allied Chemicals on coal byproducts. Soon these firms were developing new products based on other, but closely related, chemical technologies.

So, too, the pharmaceutical revolution of the 1940s – the production of sulpha, penicillin and other antibiotics that transformed the American pharmaceutical industry – was carried out not by new start-up firms, but by established companies such as Merck (the US descendant of the German company), Abbott Laboratories, Squibb, Upjohn, Bristol, Lederle (a subsidiary of American Cyanamid), Parke, Davis and other long-established enterprises. In that revolution, chemical processes replaced mixing in production; selling prescription drugs to doctors and hospitals became more important than selling over-the-counter branded packaged products in marketing, and firms' development organizations became much larger and more costly (Temin, 1980, pp. 75–80). In very recent years the same firms, along with Du Pont and other chemical companies, have become active in the new biotechnology industries. In chemicals and pharmaceuticals far more new products were commercialized (and even new industries created) by established firms that had institutionalized the learning processes than by small entrepreneurial start-ups.

In the electrical and electronic industries, the institutionalizing of product development had much the same results as in chemical and pharmaceuticals. One difference was that the four global first movers began by producing a wider variety of products than did those in the chemical industries. Before the turn of the century, GE, Westinghouse, Siemens and AEG were making and selling a full line of equipment for electrical power, light and urban traction, as well as smaller motors for industrial power. All four quickly institutionalized the learning acquired through the development of their initial products. Soon both German and American companies moved into the production of electric locomotives for long-distance rail travel. They developed and then marketed electric heaters, stoves, refrigerators and other consumer appliances. Westinghouse moved into lifts, becoming the first major challenger to Otis Elevator. GE's development work in improving light bulb filaments brought it into the production of alloys; its efforts in improving installation materials led it into plastics and varnishes. The German electric firms concentrated more than their American counterparts on the production of electrochemicals, railway and other signals, electric cables, and telephone and telegraph equipment. (Siemens had been the leading producer of telegraph equipment in Europe before the coming of electrical revolution.) The leaders in both countries commercialized X-ray and other medical equipment, and measuring and other sophisticated instruments (Chandler, 1990a, pp. 217–21, 472–3, 543–8).

Their development work in these latter fields, particularly that on the light bulb, led these firms to create the radio and then television industries. A British enterprise had been formed to produce Marconi's invention, but its improvement and commercialization was carried on primarily at GE, Westinghouse and Western Electric (the manufacturing subsidiary of AT&T). In 1919 these three firms and the American subsidiary of British Marconi formed the Radio Corporation of America (RCA). RCA soon began to distribute radios manufactured primarily by GE and Westinghouse, and then it took over their radio manufacturing as well as selling. In 1932, when RCA was spun off from its parents, its development department began to concentrate on commercializing television. Much earlier in 1903, Siemens and AEG had formed Telefunken – a fifty–fifty joint venture – to develop 'wireless telegraph'. During the 1920s and 1930s it began to commercialize radios in Europe and then to investigate television in the same manner as RCA did in the USA (Aitken, 1985, pp. 13, 25, 67, 232–3, 434–5, 457, 465, 586 and Ch. 10; Chandler, 1990a, p. 466).

If the chemical and electrical equipment industries were those that drove economic growth and profoundly transformed the ways of production, distribution and everyday life in the early years of this century, certainly the electronic information system industries have played much the same role in recent years. The commercializing of computers, semiconductors and other components of the information systems industry only differed from those of the early industries in detail. Entrepreneurial start-up firms played more of a role (though a less significant one than the conventional mythology suggests) than they did in the commercializing of radio and television. The mainframe computers – the first type of computers to be widely used – were commercialized almost wholly by long-established companies. In Europe these were producers of electrical and telephone equipment. In the USA they were business machinery companies. The American story emphasizes the technology/market interaction so critical in the learning involved in new product development. Entrepreneurial start-ups continued to play a significant role only if they made the three-pronged investment essential to exploit the economies of scale and scope (Cortada, 1987; Flamm, 1988, Chs 4–6).

The computer came out of universities. Its initial funding came from the government, largely from the Department of Defense, its first significant market. But the three-pronged investment essential to produce the new machines at a cost low enough to exploit the potentially huge civilian market – that of industry, commerce, transportation and finance – were made by business firms. Of the six major producers of the mainframe computer, only one, Control Data, was an entrepreneurial enterprise. The first to commercialize the mainframe computer was Remington Rand (Remington had been a first mover in typewriters in the 1880s) with its introduction of UNIVAC in 1951. In 1955 Remington Rand merged with Sperry to obtain greater technological capabilities. Burroughs Adding Machine (also its industry's first mover in the 1880s), National Cash Register (again the first mover in its industry in the same decade) and Honeywell (a heat-control equipment company which purchased a calculator firm in the 1930s to compete with International Business Machines, or IBM), and IBM itself followed suit. These firms had sales forces in place. They acquired (either through acquisition or direct investment) the necessary technological skills and then expanded their managerial force to handle the new product line.

While these six pioneered in the initial development of the mainframe, it was IBM which engineered a massive three-pronged investment that made it the first mover in the new industry. The strategy of its top managers, particularly Thomas Watson, Jr, was to reach as wide a commercial market as possible by utilizing the cost advantages of scale and scope. They hoped, through standardization and compatibility of machines and components, to produce a broad line of mainframe computers with peripherals for a wide range of uses. It took several years of intensive research in product design, in markets and in the development of production processes before IBM's System 360 came on stream. Then IBM's continuing impressive investments in research and production, the swift expansion of its international marketing organization and a rapid increase in its management ranks gave the company the dominant position it still has today (Sobel, 1981, Ch. 10).

With the single exception of Control Data, IBM's successful mainframe competitors continued to be established business machine companies which acquired electronics companies to improve their production and research competencies. In contrast, major electronics companies dropped out of the business. Raytheon and GE sold their operations to Honeywell; RCA's computer activities were acquired by Sperry Rand; and Philco dropped its computer operations soon after it was taken over by the Ford Motor Company. These companies had as great a potential for success in computers as the business machine companies did. They had long experience in product development but, by the 1960s, they had become widely diversified. Because computers were only one of many product lines, their senior executives were unwilling to allocate the time and make the large investments necessary to build an effective competitive capability. (Much the same thing has happened more recently in consumer electronics.)

Entrepreneurial start-ups did play a critical role in designing machines that used different technologies for markets other than those of the mainframe. At Digital Equipment (DEC), Kenneth Olsen's large investment in manufacturing the PDP-8 line, accompanied by the creation of a worldwide marketing network and a sharp rise in the number of managers, made DEC a first mover in minicomputers. Edson de Castro, the engineer who headed the design team for the PDP-8, made a comparable set of investments when he left DEC in 1968 to form Data General. However, a third pioneer in microcomputers, Scientific Data Systems, failed to scale up. It quickly disappeared from the scene after it was acquired by Xerox in 1969.

The most successful challengers to DEC and Data General, however, were not other entrepreneurial start-up enterprises but established managerial companies. By 1980, DEC ranked second and Data General fourth in revenues generated. IBM was first, Burroughs third, and Hewlett-Packard, an established producer of electronic measuring and testing instruments, fifth. The sixth was Wang Laboratories, a first mover with a new product for a different market: word processing and office systems. Together these six accounted for 75 per cent of the revenues generated in the minicomputer branch of the industry (Archbold, 1983, p. 92).

Much the same pattern appeared in personal (micro) computers. By 1980, the first entrepreneurial start-up companies to make extensive three-pronged investments – Apple Computer, Tandy and Commodore – accounted for 72 per cent of US dollar sales. (These three quickly drove out the three pioneers that had accounted for 50 per cent of sales in 1976.) Two years later, however, three established companies – IBM, Hewlett-Packard and the Japanese company, NEC – moved in and captured 35 per cent of the market, driving the entrepreneurial first movers' share down to 48 per cent (Chandler, 1990a, pp. 612–13).

All these firms became multinational: IBM was almost immediately the leading producer of mainframe computers in Europe; DEC led in microcomputers; and by the mid-1980s, Apple and IBM were already world leaders in personal computers. Abroad they competed with long-established European and Japanese enterprises. True to form, British pioneers failed to make the necessary investments. By 1974, only a little over a quarter of all computer installations in Britain came from British producers.

Major players in the computer industry have changed little in the past decade. Of the 20 largest hardware producers in 1987, only two were founded in the 1980s. The single successful challenger in existing sectors was Compaq (ranked fourteenth in total revenues), whose management announced in its very first annual report that it thought of itself 'as a major company in its formative stages rather than as a small company with big plans' and invested accordingly (Compaq Computer Corporation, 1983). The other new company, Sun Microsystems (ranked twentieth), followed the path entrepreneurial companies had traditionally taken and developed a new architecture for a new market, work stations.

These US computer companies continued to be powerful competitors globally, even as they encountered hurdles like the rising cost of capital, the fluctuating dollar, and anti-trust and other regulatory legislations that are often cited to explain the decline of other industries. In 1986, they still enjoyed just under 60 per cent of the European market in mainframe and minicomputers, and just over 20 per cent of the Japanese market. In Europe, IBM's market share was 35 per cent, DEC's 7 per cent, Unisys's (formed by the merger of Burroughs and Sperry Rand in 1986) 5 per cent, and Hewlett-Packard's 3 per cent. In Japan, IBM has 15 per cent, Unisys 3 per cent, and NCR 2 per cent (Flamm, 1988, pp. 168, 201). In microcomputers, Apple and IBM (with Japan's NEC) accounted for 50 per cent of the world market. Foreign competition in the USA remained limited, except in some peripherals.

In semiconductors, the story has been very different. This industry, which supplies critical components for computers, telecommunications, factory automation, robotics, aerospace and production controls, was created in the USA. In the mid-1970s, the pioneering American companies held 60 per cent of the world market, 95 per cent of the domestic market, half the European market, and a quarter of the Japanese market. By 1987, their world market share had fallen to 40 per cent, while the Japanese share had risen to 50 per cent. The USA had become a net importer, with the Japanese supplying 25 per cent of its market. Japanese enterprises controlled over 80 per cent of the world's sales of DRAMs (Dynamic Random Access Memory) commercialized by the US company, Intel. By the late 1980s IBM, Intel, and a small number of other producers of semiconductors were working with the Defense Department through Sematech to try to save the industry (Anchordoguy, 1989, pp. 136–48; Dertouzos, Lester and Solow, 1989, pp. 248–64; Chandler, 1990b).

What happened? Again, the highly diversified electronics companies with the greatest capabilities for production and continuing research in semiconductors – RCA and GE – pulled out, while Ford's take-over of Philco destroyed the potential there. More serious, though, was the pioneering companies' failure to institutionalize the learning processes within their enterprises. Instead of making the long-term investments essential to the creation of organizational capabilities and then continuing to reinvest, these entrepreneurial firms remained small or sold out, often to the Japanese. Repeatedly, groups of engineers left their companies to start new ones, often lured away by venture capitalists anxious to finance still another start-up. Their departure, in turn, destroyed existing teams who, through learning by doing, were creating the next generation of improved products in this technologically dynamic industry.

More serious than the current swift loss of market share has been this continuing failure to institutionalize the learning process. The production of semiconductors has become increasingly capital-intensive. The state of the art plant which cost $150 million in the mid-1980s in the early 1990s will now be over $300 million. Development costs have increased accordingly.

In Japan, the makers of semiconductors all had long experience in the development of new products and processes. Like IBM, all made computers. Unlike IBM, they produced semiconductors not only for themselves but also for the larger domestic and international markets. These companies – NEC, Hitachi, Toshiba, Mitsubishi Electric, Oki Electric, Fujitsu and Matsushita – had been producers of electrical and telecommunications equipment since the 1920s and earlier. They had diversified but, unlike their American counterparts, only into closely related product lines. Moreover, all belonged to *keiretsu*: groups of allied, independent enterprises with their own banks and trading companies. These allies provided further financial, marketing and research benefits. By aggressively exploiting the cost advantages of scale and scope, these Japanese companies easily drove most entrepreneurial American competitors and the large, more widely diversified electronics companies out of the business or into highly specialized niches.

CONCLUSION: LEARNING, TECHNOLOGICAL CHANGE AND THE BUSINESS FIRM

In the past century technological change in chemical, electrical and electronic products has transformed the ways in which we work and live. Such change has revolutionized urban living, rural agriculture, household work, the processes of production, distribution, transportation and finance, medicine and medical practices, communications of all kinds, and even entertainment. The business firm has played a critical role in these transformations. First it became the instrument by which innovations were commercialized. Within a new firm inventions were designed and redesigned and processes of production shaped and reshaped to make a product whose performance and cost would bring about its widespread use. Then the lessons learned in the effort to commercialize the new product were incorporated and routinized in an internal development organization. Not all firms in the new capital-intensive, transforming industries played this role, as the British experience emphasizes. Only those firms that learned how to integrate technological innovation with market needs and then institutionalized the learning process became and remained long-term players in their industries.

It was the institutionalizing of the learning involved in product and process development that gave established managerial firms advantages over entrepreneurial start-ups in the commercializing of technological innovations. Development remained a complex process involving a wide variety of usually highly product-specific skills, experience and information. It required a close interaction between functional specialists, such as designers, engineers, production managers, marketers and managers, who were knowledgeable about sources of supply. Such individuals had to coordinate their activities, particularly during the scaling-up processes and the initial introduction of the new product to the market. The process was costly, time consuming and risky. In financing new product development, existing firms with established core lines had retained earnings as a source of inexpensive capital and often had specialized organizational and technical competence not available to new entrepreneurial firms.

For similar reasons the established firm that developed products close to its core technologies had advantages over more widely diversified enterprises, particularly over those companies whose managers acquired product lines in industries where they had little or no product-specific experience. Such diversification dissipated retained earnings and specialized competencies. Such firms had difficulty in competing with more focused enterprises which concentrated their resources on a smaller number of more closely related products. The success of American business machinery makers and the failure of American electronic equipment firms in the development of the mainframe computer make this point.

Again, for much the same reasons, regulated or government-owned utility companies that provided power and light, transportation companies that operated railway and airlines, and government agencies responsible for developing nuclear power (the AEC and then the Department of Energy) and for space exploration and control (NASA) have played in capitalist economies a lesser role in the technological change than business firms. The goal of enterprises in such systems has been to provide effective power or light, or transportation, or to meet defence objectives. It has rarely been to develop, produce and distribute new products and processes not directly related to those goals. The companies and agencies in these systems did become markets for, and their managers worked closely with, business firms that developed power plants, transportation and aerospace equipment, computers and other components needed in their operations, but these nonmanufacturing companies and agencies were only one of many markets for those industrial firms which acquired and then institutionalized the learning involved in technological change. For the for-profit industrial firms in rivalrous competition, the institutionalizing of such learning helped to assure present profits by improving the performance of existing products and helped to generate future profits by commercializing new goods and new methods.

For the business historian an awareness of technological change and the process of learning about such change is essential to the understanding and evolution of the functions, activities, role and performance of that discipline's primary unit of study: the firm. Indeed, the history of modern business and the history of modern technology are intimately intertwined. Just as business historians cannot comprehend many aspects of business history without a knowledge of the intricacies of technological change involved in the processes of production and distribution, so historians of technology need to understand business activities, procedures, organizations, objectives and methods to comprehend fully the processes of technological change, particularly the place of learning involved in that change. Such knowledge can help to explain why technological change took the form it did, when it did, where it did and in the manner that it did. Together the study of technology and business and their interaction in capitalist economies is essential to understanding broader economic change, and in particular the recent transformation of much of the world's economies from agrarian, commercial and rural to industrial, urban and technologically dynamic.

References

Aitken, H.G. (1985) *The Continuous Wave: Technology and American Radio, 1900–1932* (Princeton, NJ: Princeton University Press).

Anchordoguy, M. (1989) *Computers, Inc.: Japan's Growing Challenge to IBM* (Boston, Mass.: Harvard Business School Press).

Archbold, P. (1983) 'The Datamation 100: Welcome to the Club', *Datamation* (June).

Carlson, B. (1989) 'Elihu Thomson. The Thomson-Houston Electric Light Company, and the Development of Electric Light', a paper delivered at the Business History Seminar, 13 March.

Chandler, A.D. Jr (1990a) *Scale and Scope: The Dynamics of Industrial Capitalism* (Cambridge, Mass.: Belknap Press).

—— (1990b) 'The Enduring Logic of Industrial Success', *Harvard Business Review*, 90 (March–April).

Cortada, J. (1987) *Historical Dictionary of Data Processing* (New York: Greenwood Press).

Dertouzos, M.L., Lester, R.K. and Solow, R.M. (1989) *Made in America: Regaining the Productive Edge* (Cambridge, Mass.: MIT Press).

Flamm, K. (1988) *Creating the Computer: Government, Industry and High Technology* (Washington, DC: Brookings Institution).

Haber, L.F. (1971) *The Chemical Industrial, 1900–1930, International Growth and Technological Change* (Oxford University Press).

Hounshell, D. and Smith, J.K. Jr (1988) *Science and Corporate Strategy: Du Pont R&D, 1902–1980* (Cambridge University Press).

Nevins, A. (1954) *Ford: The Times, the Men, and the Company* (New York: Scribner's).

Reader, W.J. (1970) *Imperial Chemical Industries: A History. Volume I. The Forerunners, 1870–1926* (Oxford University Press).

Rosenberg, N. (1979) 'Technological Interdependence in the American Economy', *Technology and Culture* (January).

Schumpeter, J. (1942) *Capitalism, Socialism, and Democracy* (New York: Harper & Row).

Sobel, R. (1981) *I.B.M.: Colossus in Transition* (New York: Times Books).

Temin, P. (1980) *Taking Your Medicine: Drug Regulation in the United States* (Cambridge, Mass.: Harvard University Press).

4 Nineteenth-Century American Patent Management as an Invisible College of Technology

Carolyn C. Cooper

History of technology in the post-medieval Western world has to be a history that deals with dynamic change, one that attempts to explain how positive feedback took place among increasingly integrated material and social systems, so that technological change became not an occasional, but an incessant, event. Since the concept of intellectual property protected by patents for invention emerged in history at the same time as this increasing dynamism of technology, it is tempting to assume there is some connection between the two.[1] On that assumption, students of technological change have frequently viewed patents as incentives to invent or as a measure of invention. This chapter does not debate the validity of those views, but attempts to outline quite a different feature of a patent system in action: as an institution it elicits behaviour aimed at patent management, which in turn acts as an invisible college of technology, in which learning stimulates further technological change.

If one reads the various kinds of American patent record of the nineteenth century between the lines, so to speak, they reveal an unexpected internal dynamic: a social system of human interactions required for patent management which focuses on communication of knowledge, and in turn generates new knowledge.[2] Patent management in this country in the first several decades of the nineteenth century appears to have been important – arguably more important than other more explicitly educational institutions at that time – for the teaching and learning of new technology. This chapter will explain how those interactions created and diffused new knowledge among the participants in the system. That was a period of rapid flowering of mechanical technology in the USA; accordingly, the examples given here are mechanical, rather than, say, electrical or chemical.

PATENT MANAGEMENT

Patent management by inventors (or patentees) brought them into interaction with certain other persons in several situations, each of which had another ostensible purpose, but also served as an occasion for technical teaching and learning. These were: (a) *applying for a patent*, when, in order to become patentees, inventors dealt with patent agents, model builders, financial backers, patent examiners, and (in cases of interference) with other inventors and *their* agents; (b) *using, selling, or licensing* the patent, in which the patentees interacted with partners, financial backers, assignees and licensees, any of whom might also be users of the invention; (c) *defending the patent from infringement*, when patentees communicated not only with any of the above, but also with other inventors, lawyers, judges, 'expert' witnesses and juries. Sometimes patentees wanted (d) to *obtain a reissue of the patent* or (e) to *extend its term*, which would bring them additionally into communication on technical matters with patent office commissioners and/or US congressmen of both houses. All these activities by inventor/patentees, financial backers, assignees, patent examiners, judges and so forth were patent management activities; as a whole they constituted the patent management system.

Patent management thus harnessed economic processes with legal, governmental and ultimately intellectual processes in defining not only who temporarily 'owned' which inventions, but also what features of technique were important in distinguishing one mechanical object from another. Patent management interactions in these several arenas determined (and sometimes redetermined) what was new and different about the putative invention that made it deserving of a patent: that is, the right for a number of years to regard it as one's private property, from which others could be legally excluded. All those participants in the system had occasionally to compare two items of technology at a time, and decide either yes, gizmo *x* was significantly different from gizmo *y*, or no, gizmo *x* and *y* were substantially the same. Inevitably, there were disagreements about perceived difference and similarity; then the participants could argue with one another about their perceptions in the arenas provided by the system. Successful litigation vindicated some of those acts of perception. From July 1836, when examiners began to scrutinize patent applications not only for form but also for originality of content, the granting of patents also vindicated some perceptions.[3] Over time, these vindicated perceptions amounted to the social construction both of particular inventions and of the decision rules for 'invention' in general (Cooper, 1987b, 1991a, 1991b).

INVISIBLE COLLEGES

In communication and disputation with one another about the mechanical technology that was developing before their eyes, the various participants in the patent management system constituted an 'invisible college', or rather, a congeries of 'invisible colleges'. This term was first applied to the nascent group of scientific amateurs in England who communicated informally with one another before they organized and became 'visible' as the Royal Society in 1662 (Merton, 1988). Then their informal communications by letter were supplemented by recorded minutes of the Royal Society, once it began having regular meetings at which experiments were demonstrated and discussed.[4] Members of the invisible college probably also sometimes met face-to-face and simply talked, but almost by definition we have no evidence of what they discussed.[5] However, their letters to one other served 'to gain an appreciative audience for their work, to secure priority and to keep informed of work being done elsewhere by others' (Price and Beaver, 1966, p. 1011).

The phenomenon of the invisible college has persisted in science even while more organized institutions, such as societies, academies, universities and research laboratories, have taken on the functions of publishing scientific results. In the 1960s it became the subject of study by sociologists of science (Crane, 1972; Chubin, 1983). Through indices of scientific journal citations, Derek de Solla Price and others identified geographically widespread modern-day invisible colleges centred on particular research topics, whose members communicated among one another 'in some informal fashion behind the backs of the conventional journals' in order to stay abreast of rapid developments in that specialty (Price, 1971, p. 75). In the context of modern, increasingly subdivided 'Big Science', some invisible colleges have become visible enough to have names like 'the Phage Group' or 'Information Exchange Group No. 1 on Oxidative Phosphorylation and Terminal Electron Transport', and even to have logistical support (though not control) by a regular scientific organization, such as the National Institutes of Health.[6]

This modern type of invisible college in science has been called an 'elite' who 'automatically reinforce their exclusiveness only by their excellence and by the overpowering effect of their contribution relative to that of the rest' of the scientists (Price, 1971, p. 75). They 'meet in select conferences (usually held in rather pleasant places), they commute between one center and another, and they circulate preprints and reprints to each other and they collaborate in research' (Price and Beaver, 1966, p. 1011). The invisible colleges are by hypothesis of finite duration: they reach a stage of exhaustion after solution of the major problems in a specialty, anomalous evidence provokes increased questioning of the operating paradigm, and presumably new colleges form around new specialties.[7] According to some students of the phenomenon, the written output of invisible colleges grows exponentially, and can be plotted over time in S-shaped logistical curves of varying steepness, reflecting this life cycle as colleges die and ever-increasing numbers of new ones form (Crane, 1972, pp. 173–86).

Students of informal scientific groupings have continued to coin new terms for them in addition to 'invisible college' – for example, 'social circle', 'community', 'network', 'specialty' – but 'none has enjoyed widespread usage . . . the meaning changes with the . . . user' (Chubin, 1983, p. 8). One student has usefully characterized a social circle in contrast to an organized group as a 'network' having '(1) no clear boundaries; (2) indirect interaction (not everyone has to know everyone else or have contact with everyone else); (3) . . . no formal leadership; (4) . . . [no] instituted structures or norms . . .; and finally (5) [a tendency] to be pegged or draped around other structures' (Bystryn, 1981, p. 120). These largely negative characteristics would fit all sorts of invisible colleges. Their positive common denominator, however, is that their members generate and share not just behaviour, but knowledge.

KNOWLEDGE AND COMMUNICATION IN SCIENCE AND TECHNOLOGY

The product of these invisible colleges of scientific communication *is* science, in its written form. The diffusion of knowledge in a college stimulates creation of new knowledge, which is in turn diffused. But technical knowledge is notoriously less communicable by the written word than is scientific knowledge (Ferguson, 1977). It can be written *about*, but not directly *written*.[8] Although it clearly shares with science the characteristic of phenomenal growth and proliferation in recent centuries, technology appears primarily in the form of artefacts and processes, and only secondarily, if at all, in the form of writing. Derek de Solla Price pointed to the contrasting motivations of scientists to claim intellectual property by prompt publication and of technologists to delay communication until securing a patent, and summed it up by calling science 'a cumulating activity which is *papyrocentric*, while technology also cumulates, but in a *papyrophobic* fashion' (Price, 1965, p. 561).

Despite a relative lack of explicit written product, however, the patent management system of pre-Civil War USA did resemble invisible colleges in several other respects. First, it *was* invisible, perhaps even to its own members, who identified themselves in terms of other roles – as lawyer, practical mechanic, inventor, engineer, patent agent, merchant, and so on – rather than as 'technologists'. So it did tend to be 'draped around' other social structures. In addition, for the most part it was self-selecting, without a formal entrance test for membership. Also, they met irregularly and fleetingly in small face-to-face groups – in workshops, factories, lawyers' chambers, courtrooms and the offices of businessmen, patent agents or examiners – to communicate their perceptions of similarities and differences between different inventions. They did not all know one another. But yes, on some occasions they did write to one another, or their statements were taken down in writing and sent to be read by and to other participants, in the course of drafting a patent application, selling patent assignments or licences, or conducting interference or infringement suits.[9]

In order to proceed, however, the members of an invisible college of technology had to augment verbal by non-verbal means of communication. They could not just write letters: they also drew pictures of the technical devices in question; they made three-dimensional models of them; and they pointed to samples of machines' products as well as to the machines themselves. Non-verbal communication was important in all the patent management arenas mentioned above, where members of a technological 'invisible college' met to share their technical perceptions. The next four sections of this chapter illustrate how patent management behaviour in these arenas created 'knowledge'.

PATENT APPLICATION

From its beginning in 1790 the rules of the patent system called for drawings and models to accompany patent applications. After the totally destructive patent office fire in December 1836, models and drawings were 'restored' to the record, as were written specifications. Patent examiners in the reformed Patent Office consulted a growing collection of all three modes of description in deciding whether the claimed features of an invention were truly original as well as useful (see Figure 4.1). After the patent examiners had 'hands on' experience of them, the models of patented devices were also on public view; when housed from the 1840s in glass cases in the 'Model Hall' of the newly-built Patent Office building, they became a popular tourist attraction in their own right.[10]

Outside the Patent Office, the occupational specialties emerged of patent agent and patent attorney, to help inventors obtain and manage patents. For a fee, they would advise inventors as to the patentability of their inventions, help them write patent specifications, and represent them in dealing with the Patent Office and in lawsuits. Patent agents and agencies such as Munn & Co., associated with the publication *Scientific American*, were located in New York, Washington, DC, and other large and small towns.[11] By 1860 there were nearly three dozen patent agencies in Washington, DC, alone and by 1870 Munn & Co. was boasting that they had handled nearly 20 000 patents in over 20 years of experience (Munn & Co., 1871, p. 17; Post, 1976, p. 160).

Patent agents and would-be patentees also looked at the models and drawings at the Patent Office and learned from them.[12] In part, the models served to reveal how already-patented devices worked so that other inventors could avoid wasting effort on them, or could decide to 'invent around' them. James Watt, for instance, had invented 'around' the crank in devising the sun-and-planet gear for his double-acting steam engine (Robinson and Musson, 1969, pp. 89–95 and plates). In part they also served as a convenient reference library of mechanical motions and linkages that might be used in other combinations for other purposes. As such, they were similar in function to the three-dimensional 'mechanical alphabet' that Swedish engineer Christopher Polhem made for instructional purposes in the eighteenth century (Johnson, 1963, p. 158; Ferguson, 1977, p. 835). One early visitor to the Patent Office said the models 'were of incalculable value to the real mechanician', and estimated that his visit would enable him to 'reduce the price of spinning wool, from eight, to one cent per pound' (Ferguson, 1977, p. 836).

The very distinction between elemental bits of technique and variable 'combinations' of them was part of the learning that participants in the

Figure 4.1 Patent examiners at work, 1869

One is consulting the models and drawings for patented ploughs.

Source: *Harper's Weekly Magazine*, 10 July 1869, p. 445.

patent management system could gain in this exercise. For instance, in 1813 Thomas Jefferson joined those who expressed doubt that the ancient chain-of-pots and Archimedean screw should be patentable when Oliver Evans used them in gristmills to move grain and flour instead of water (Ferguson, 1980, pp. 57–8). But after persons in the patent management system paid attention to such combinatorial possibilities as in Evans's gristmill, new combinations of old elements became routinely patentable. One may suppose that practice in recognizing the 'newness' of combinations of old elements was important in promoting cumulativity in technology and particularly patented technology, for it expanded the concept of (patentable) 'invention'.

The patent office itself was tardy in taking on the responsibility of making information freely available concerning patents that were currently in force. Before this kind of information was published, inventors had to rely on correspondence, hearsay and visits in person to find out how other machines actually worked. David Wilkinson, the American inventor of the slide-rest lathe for cutting screw threads, recalled in 1846 that 50 years earlier he had 'made a model in miniature and had thought of trying to procure a patent, but was afraid there might be something somewhere to interfere with me, already in use. So I started off [from Pawtucket, Rhode Island] to make inquiries'. He went to New York but found only old methods of screw-cutting. Proceeding from there up the Hudson by sloop for five days to Fishkill, he 'went ashore and walked some thirty miles to Canaan', Connecticut, to check up on a rumoured innovation there. Then he

> returned to New York, and from there went to Philadelphia, and found no screws made there except after the same mode as in New York. I heard of screws being made on the Brandywine, but my informant assured me they were made the same [old] way . . . I now returned home, and in the year 1797, went again to Philadelphia, when Congress was in session, and made application for a patent. (Wilkinson, 1846, p. 88)

Had he travelled as far as London, Wilkinson might have found that the English machine tool builder, Henry Maudslay, was simultaneously developing a slide-rest lathe!

Once descriptions and illustrations of patented devices were published, they did not totally eliminate the need for such journeys as Wilkinson's, but they immensely eased the task for would-be inventors of keeping themselves informed. Despite the long-standing British example of publishing patent descriptions in *The Repertory of Arts & Manufactures*, Dr William Thornton, the first patent office superintendent (1802–28), was unwilling to allow such publication, feeling more keenly a responsibility to protect the intellectual property of the patentees from public gaze. In the 1820s, however, the newly organized Franklin Institute of Philadelphia fought with him and won the right to publish descriptions of patented devices (Sinclair, 1974, pp. 42–5). Thomas P. Jones, editor of the monthly *Journal of the Franklin Institute*, made patent abstracts a regular feature from 1826 and included some lithographic illustrations. Two decades later, the weekly *Scientific American* began publication and included patent listings as a regular feature, plus selected descriptions and pictures. Only in 1843 did the Patent Office itself begin including extracts from new patents in its *Annual Reports*, and in 1853 it started showing drawings. One could hire a copyist to write out a whole specification, but otherwise a trip to the Patent Office was still necessary to inspect a patent in detail. In 1872 the Patent Office began printing multiple copies of the complete specifications of new patents (including illustrations by photolithography), made them available for a modest fee, and began issuing weekly Patent Office Gazettes summarizing them. Thus, during the course of the nineteenth century formerly haphazard exchanges of information within the invisible college of technology were published with increasing frequency and completeness; first at private initiative and expense, then as a governmental responsibility.

Publication of patent information, of course, advertized the invention to a geographically wider pool of potential buyers of assignments and licences. Although it also made piracy easier, this publicity worked to the inventor's advantage in marketing his patent, helping to counteract the disadvantage of losing his secrecy. From the point of view of promoting development of the nation's economy and technology, it was of course helpful to have patent information disseminated widely. Even during the period in which a patent restricted use of a particular combination of technical information, such dissemination would promote general technical sophistication among mechanics and inventors.

The Franklin's *Journal* editor, Jones, moved his office to Washington, DC, in 1828 and became the second superintendent of the patent office for a year. He was a patent agent intermittently from 1836 and served two years as patent examiner in the reorganized patent office, all the while continuing to edit the *Journal* until his death in 1848 (Sinclair, 1974, pp. 198–200). Other patent examiners also sometimes became patent agents and vice versa (Post, 1976, *passim*). In both roles they had to communicate with one another and with inventors in an invisible college of patent management. When inventors or their agents, in applying for a patent, talked and wrote to patent examiners about the patentability of different features of their inventions, they were engaged not only in patent management but also in teaching and learning mechanical technology. Their correspondence, preserved in the Patent Office files of patented inventions, has been likened to the negotiations between editor and writer about revisions before publication.[13] Puzzling gaps in the recorded negotiations suggest that unrecorded face-to-face conversations also took place between the examiners and the inventors or their agents. Eventually either the patent was granted or the applicant gave up (Reingold, 1960, p. 160). In both cases, knowledge had been shared and increased.

USING, SELLING AND LICENSING A PATENT

Even before obtaining a patent, an early nineteenth-century inventor would frequently market it prospectively, to seek financial backing for the expenses of R&D and of the patent application. An inventor sometimes turned to his local congressman to fill the roles both of financial backer and patent agent in applying for the patent. For instance, planing machine inventor William Woodworth pre-assigned half his patent right to congressman James Strong of Columbia County, New York, who put up $1500 in 1828 for 'building a model and first machine and for the taking out of Patent'. (The patent fee itself was only $30.) Their subsequent 'expenditures and loss in further experimenting and putting up and running' two machines in New York City cost $8000 more.[14] Like Woodworth and

Strong, an inventor and backer(s) commonly shared ownership of a machine patent, and could then proceed to produce machines for sale, to use the machines to make other goods, to sell patent rights to others, or to license use of the machines.

As property that was intangible, a patent could be divided geographically; some patents also lent themselves to division by the various purposes for which the patented device (a machine, say) could be used. A patented invention could therefore be sold or leased in different geographical territories for the same purpose, or for different purposes in the same territory.[15] This practice expanded the potential pool of investors on which an inventor could draw for financial support. The geographically scattered assignees and licensees became part of an 'invisible college' focused on the specific technology expressed in the patent, and of somewhat larger 'colleges' concerned with related technologies.

These multiple assignments and licences of a given patented invention by purpose and by territory speeded the diffusion of that invention without having to wait for the actual production and shipment of the invented device. An important desideratum in writing and illustrating a patent specification was that it be intelligible to anyone 'versed in the art'. To the extent that this legal requirement was in fact fulfilled, it enabled users who were so versed to build their own machines. In the case of fairly heavy and bulky machinery, it was probably often much easier in the early decades of the century for a skilled would-be user at some distance to pay for a machine's designs and build it himself with local materials and labour than to have a patentee or assignee build and ship an actual machine. For instance, the Blanchard Gunstock Turning Factory sold a lastmaking lathe for about $400 (plus $100 annual royalty) in the mid-1820s; the cost of a user-built one in the mid-1830s was estimated at $125–75 (Cooper, 1991a, pp. 173–4). Once a machine was in operation, still other local mechanics could observe and learn from it. Thus diffusion of machines could take place through the postal system without waiting for the full emergence of the machine industry.

Non-verbal means of communication were used in marketing the patent, as well as applying for it. Late in the century, at least one patent agent even went on tour with a wagon exhibiting models and specifications in order to drum up business in rural areas (see Figure 4.2). In a typical earlier instance, Thomas Blanchard, inventor of the irregular turning lathe in 1819, displayed his own small working model of it that summer and autumn at taverns in his home town of Millbury, Massachusetts, and in Boston, where potential backers came to view it. Some of them formed the Blanchard Gunstock Turning Factory,[16] which sold assignments and licences for Blanchard's 1820 patent for making tool handles and shoe lasts in various territories. Once the network of Blanchard's lastmaking assignees and licensees began using the machine, they acted as an 'invisible college', exchanging information (presumably by visits and hearsay) on how to manipulate it so that it would turn several well-proportioned sizes of last from a single model. In the mid-1930s Collins Stevens, a licensee in Boston,

OUR EXHIBITION WAGON,
Wm A. BELL & CO PATENT BROKERS & SOLICITORS
OF WHICH THE ABOVE IS AN ACCURATE REPRESENTATION.
ATTENTION! READ WITH CARE!
Have Your Patent Placed for Sale in Our Agency at Once.
MAKE MONEY OUT OF YOUR PATENT WHILE YOU HAVE AN OPPORTUNITY.

Figure 4.2 Harnessing visual communication for marketing patents

In 1891 William A. Bell & Co., patent brokers and solicitors of Sigourney, Iowa, outfitted this wagon for mobilized display of up to 150 patent models and copies of patent specifications. It toured mid-Western states in the summer and autumn, southern states the following winter. Bell & Co. told inventors that the wagon would appear at state fairs and other large expositions and also 'call on Manufacturers, Speculators and Buyers . . . Do not miss this good opportunity of having your patent brought before those most likely to buy territory.'

Source: From Warshaw Collection, National Museum of American History.

embodied this improvement in a fixture, called a 'fan board and lever', that the others also added to their lathes.[17] Left unpatented, it came to be regarded as an integral part of the Blanchard patent, and was eventually included in commercially-built Blanchard lathes (Cooper, 1991a, pp. 177–8). This sort of 'learning by doing', it can be assumed, also took place in assignee/licensee networks centred on other patents.[18]

PATENT LITIGATION

The more useful an invention was, the more likely there was to be infringement and therefore litigation. Giving patent examiners in 1836 the duty of determining originality of inventions did not eliminate accusations of infringement, so litigation remained a task of patent management. In individual cases, the judgement of the courts could, implicitly, overrule the judgement of the patent examiners in having granted a patent.

Non-verbal objects were used for communication in patent suits, as in patent application and patent marketing. For instance, in 1794 Eli Whitney received one of the early patents of the new republic for his cotton gin. Whitney and his partner Phineas Miller made what turned out to be a mistake by not licensing others to make and operate cotton gins. They set up their own 'ginneries' in the South in order to process the cotton and collect an in-kind fee. When Miller and Whitney failed to build enough cotton gins in New Haven to satisfy the immediately soaring demand for them, the invention was widely 'pirated' in cotton-growing southern states (Mirsky and Nevins, 1952, Chs 8–9). In the ensuing litigation to defend his patent, Whitney had to argue that the circular saw blades of a contender's cotton gin were mechanically equivalent in action to the wire-studded wooden cylinder in his gin. To demonstrate, he built a small model that had different forms of teeth rotating side-by-side to the same effect on the cotton bolls (Federico, 1960, p. 173).

Like many other patentees, Whitney lost more money in legal fees than he made from his cotton gin patent, but the point here is that his patent suit served as an occasion for study of mechanical equivalence in cotton gins by lawyers, jurors, expert witnesses and judges. Lawsuits offered an unintentional, or incidental, channel for communication of technical information, in addition to the intended channel of the patent specification, which became public property after the patent period. Court rooms and lawyers' chambers seem unlikely places for practical mechanics and engineers to hold seminars in identifying mechanical equivalents in machines, but something like a seminar was going on when such witnesses were asked to read specifications, look at models, and remember and describe machines they had seen elsewhere in forming their opinions about relevant differences and similarities in the machines.

More or less 'expert' witnesses came long distances to testify, bearing model machines and sample products of machines, or at least they sent written and pictorial depositions that were entered into the court record for response by other expert witnesses. Over time, these experiences must have helped shape and spread the criteria for mechanical equivalence that came to be accepted by the technically knowledgeable community of mechanicians in the USA. Like the members of modern invisible colleges of science who repeatedly attend conferences held in pleasant places,[19] the same witnesses frequently appeared at trials held in federal circuit courts in Boston, New York and Philadelphia. Assignees and licensees often served as witnesses for the contending patentees, but 'civil engineers', 'millwrights' and 'practical machinists' who were not 'interested' in a patent were also much called to testify, presumably for their more convincingly objective judgements. Some patent agents who were formerly examiners – such as Charles M. Keller, Samuel Cooper and Thomas P. Jones – appeared so often that they can plausibly be described as an 'elite' whose contribution was powerful. Yankee inventor Thomas Blanchard testified in so many patent trials, not all of them his own, that he lost his childhood stutter and his country bumpkin manner. His biographer wrote: 'By means of books, social intercourse in courts and elsewhere . . . his speech impediment was conquered, and he finally attained a good degree of culture and expansion of mind' (Waters, 1881, p. 260). At $2.00 a day plus expenses, other practical mechanics also travelled to attend court in distant cities where they could imbibe culture and expand their minds, while also comparing notes on technical questions.

Patent litigation involved an additional range of persons – lawyers, judges and juries – who were less practically knowledgeable already about machinery than were the mechanics, inventors and patent agents. But they, too, were required to pay attention to the evidence presented by the 'expert' witnesses and come to decisions about similarities and differences among machines. Over time, experience on such juries must have helped spread an informed understanding of mechanical matters among the general public, or at least among the taxable citizenry who composed the jury panels. Over time, repeated experience as lawyers and judges in such trials must have built up their familiarity with mechanical matters contested in the scattered federal circuit courts. Judges Story, Nelson, Kane, Woodbury and Grier, and attorneys Rufus Choate and William H. Seward, were among the mobile participants in this technological 'invisible college'.[20] Even as the patents were national in scope, so was the college.

Over time, these repeated interactions during litigation sometimes had the effect of redefining the characteristics of the machines, so that the socially constructed boundaries between them changed, sometimes dramatically. From the 1830s into the 1850s, witnesses, judges, juries and lawyers in suits against the numerous 'infringers' of the Blanchard lathe patent repeatedly undertook to compare reciprocating with circular cutters, intermittent with continuous rotation of workpieces, 'friction wheels' with 'friction points', and lengthwise with spiral paths of the cutter on the workpiece (see Figure 4.3). To do so, they looked at models of Blanchard's and contending irregular turning lathes, at patent and other drawings, and at sample products, such as lasts, from the different lathes. They also read patent specifications and listened to one another's testimony.

One persistent litigant in Philadelphia, a lastmaker named Isaac Eldridge, learned enough from such non-verbal and verbal testimony in 1848 and 1849 to reconstruct and use several variants of an 1819 'Waterbury last-making machine' that Blanchard had stated in his 1820 specification was different from his machine. By using a specifically 'different' though obsolete machine, Eldridge hoped to gain immunity from prosecution. But the judges scolded him for his 'froward ingenuity' in attempting to evade Blanchard's patent, and repeatedly imposed injunctions on him every time he rebuilt his machine with a different Waterbury-type feature.[21] In 1850 Blanchard convinced Justice Grier that intermittent lengthwise passes of Eldridge's Waterbury-type cutter along a workpiece were mechanically equivalent to the Blanchard-type continuous spiral pass, by demonstrating a

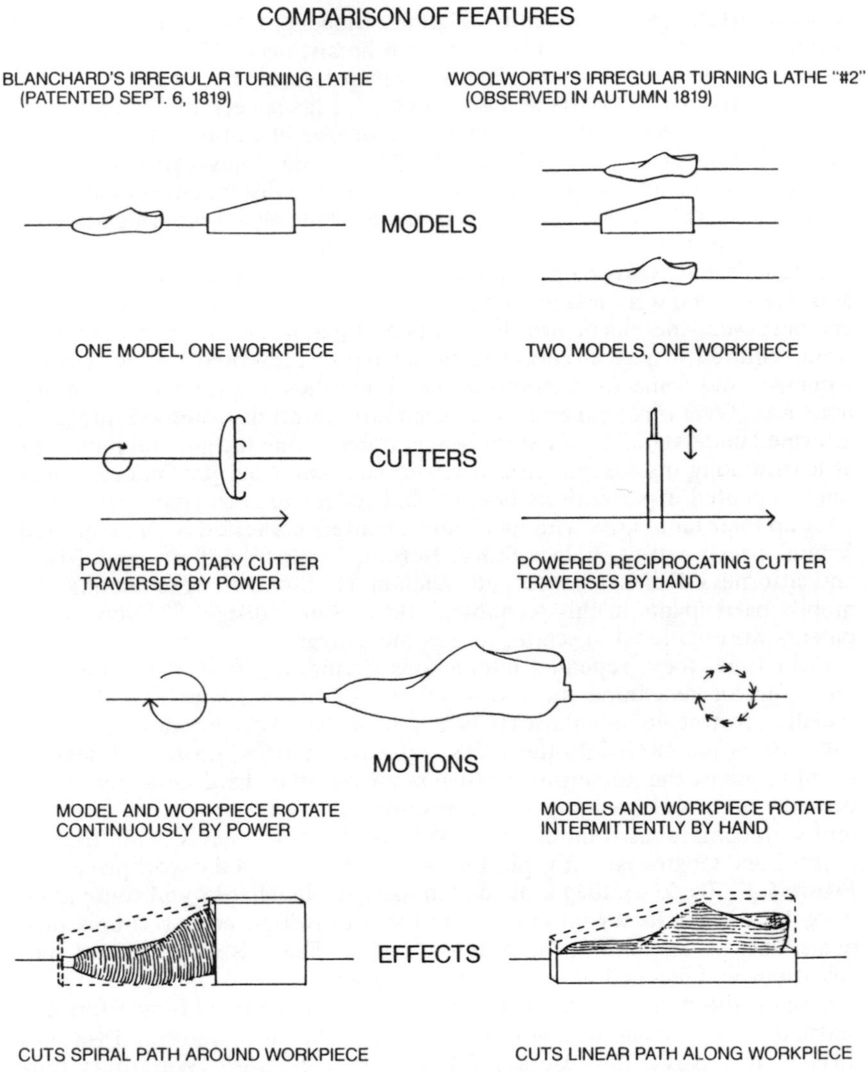

COMPARISON OF FEATURES

BLANCHARD'S IRREGULAR TURNING LATHE
(PATENTED SEPT. 6, 1819)

WOOLWORTH'S IRREGULAR TURNING LATHE "#2"
(OBSERVED IN AUTUMN 1819)

MODELS

ONE MODEL, ONE WORKPIECE

TWO MODELS, ONE WORKPIECE

CUTTERS

POWERED ROTARY CUTTER
TRAVERSES BY POWER

POWERED RECIPROCATING CUTTER
TRAVERSES BY HAND

MOTIONS

MODEL AND WORKPIECE ROTATE
CONTINUOUSLY BY POWER

MODELS AND WORKPIECE ROTATE
INTERMITTENTLY BY HAND

EFFECTS

CUTS SPIRAL PATH AROUND WORKPIECE

CUTS LINEAR PATH ALONG WORKPIECE

Figure 4.3 Irregular turning lathes compared

The Blanchard lathe and the contemporaneous 'Waterbury last-making machine' showed both similarities and differences in their modes of copying a three-dimensional model by machine. In patent disputes, witnesses, judges and juries had to study the machines and decide whether they were 'different' or 'the same' overall. (Drawings by Lyn Malone.)

model Blanchard lathe that could shift from one mode to another through a simple mechanical linkage. While winning the injunction, he thus implicitly contradicted his own patent statement of 1820. This episode shows that during three decades of patent management the Blanchard lathe had subsumed an initially distinct machine, and makes a clear example of social construction of invention.[22]

Meanwhile, a set of judges, lawyers, witnesses and so on, somewhat overlapping with the irregular turning set, also met in Boston, Philadelphia and New York in the 1840s and early 1850s during prolonged litigation over the Woodworth planing patent, to consider the various ways of planing wood by machine. The Franklin *Journal* had remarked in 1829 that the Woodworth planer, with its cylindrically mounted planing cutters, its pressure rollers and its rotary cutters for tonguing and grooving, 'differs essentially from all the planing machines which we have heretofore known' (*Journal*, 1829, p. 199). We may assume this opinion was also official, for *Journal* editor Thomas P. Jones was at this time also Superintendent of the Patent Office (Sinclair, 1974, p. 71). Other planing machines, some of them already patented in Great Britain and in the USA, had disc-mounted cutters or stationary plane irons, but defendants using such machines were brought to trial by Woodworth's son or his assignees. So the 'invisible college' met in court to compare them, feature by feature, with the Woodworth machine, using non-verbal as well as verbal evidence (Cooper, 1989).

In one trial Woodworth assignees travelled from Albany, New York, to Philadelphia to look at Daniel Barnum's machine and testified it was 'the same' as one Barnum had run in Albany, before it had been placed under perpetual injunction in a previous lawsuit there. Its planing cutter was a bevelled disc, and instead of pressure rollers to keep the board under control beneath the cutter, it had a curved iron 'fence' to bend the board slightly during its progress through the planing mill. Plaintiff Jacob P. Wilson, a Woodworth assignee in Philadelphia, argued that the fence was the functional equivalent of the pressure rollers and the bevelled disc was also 'the same as' a cylinder. In the same trial, witnesses exhibited shavings and a piece of planed board as evidence of the machines' actions.[23]

Again, the point here is not who won which trial (although the Woodworth faction predominantly did win, sometimes after an appeal to the Supreme Court), but that such experience in an 'invisible college' affected the categories in which inventors and other mechanicians recognized similarities and differences in machines. Over time, such 'learning by suing' surely increased their capacity to recognize and apply different technical means for similar ends, while readers of *Scientific American*, the Franklin *Journal* and even city newspapers could learn some things vicariously from these trials. Tuition fees for this learning – court costs and lawyers' bills – were high and caused much complaint by the litigants, but they kept coming to sessions of this invisible college, as of course did the lawyers, judges, witnesses and juries.

REISSUE AND EXTENSION OF PATENTS

Before Patent Office rules changed in 1861, patent management included the possibilities of reissuing a patent and extending its term beyond its initial 14 years. These activities brought patentees back into interaction with patent office personnel or with members of the House and Senate. Reissue was intended to correct any inaccurate or unclear wording of a specification. The patentee was allowed to withdraw the patent, reword its specification and apply to get the patent reissued for the remainder of the same term. As in the initial application proceedings, the wording of the reissued specification had to be negotiated with a patent examiner. The ostensible scope of the patent was supposed to remain unchanged in a reissue, but it nevertheless provided an opportunity, which was frequently taken, for making changes in the scope after experience – such as litigation – with the patent (Dood, 1991). Reissue could also change the terms of discourse of the 'invisible college' concerning that technology.

John C. Morris, for example, patented a machine in 1856 in which steamed wood was clamped and bent on to a form fixed in one position. Thomas Blanchard's assignees sued Morris for infringement on Blanchard's patented wood-bender of 1849, in which steamed wood was clamped, compressed endwise, and bent around a rotating form (see Figure 4.4). Morris and his assignees in turn sued the Blanchard faction (Cooper, 1991a, pp. 230–3). Both machines were capable of bending such items as plough handles and carriage wheel rims, for which there was an expanding market at the time. In 1859 and 1862, respectively, both Blanchard and Morris obtained reissues of their patents, probably intended to strengthen their arguments in the ongoing law suits.

Morris, changing the wording of his reissued specification more than Blanchard changed his, chose to de-emphasize the description of the machine itself. Instead he distinguished the action of bending from one end of the wood inward to the other (like Blanchard's) from the action of bending from the middle outward in both directions (like Morris's). He then proceeded to stake a broad claim within the second of these two 'classes' of machine which he had just defined. This had the effect of shifting the discourse in the 'invisible college' of wood-bending technique away from the structure and motions of the machines themselves, and on to the effects on their products, which was a great deal less immediately visible. A discussion of Morris and Blanchard bending methods in a major carriage-makers' journal years later still accepted Morris's distinction, even while objecting that his 'bending outward' resulted in uneven shrinkage of the wood when dying, so that wheels tended to flatten at the joints of the rims (DuBois, 1878).

Both Blanchard in 1863 and Morris in 1870 obtained 7-year extensions of their wood-bending patents from the Patent Office. To do so, any inventor had to present evidence to convince the patent commissioner that he had not obtained a fair return on his invention during the initial term of 14 years. Another way of lengthening a patent was to get Congress to pass a private

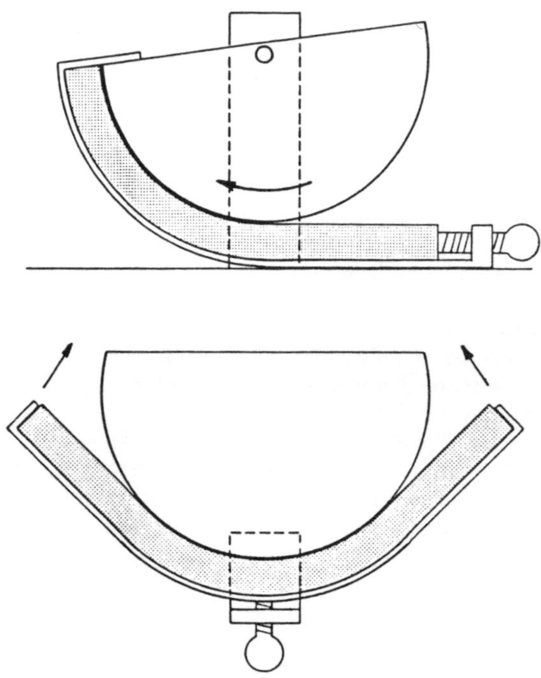

Figure 4.4 Two methods of bending wood

Above, the Thomas Blanchard method, patented 1849, with rotating form, metal band and end abutments to constrain the workpiece, and with 'upsetting' screw to apply end pressure to it. As the form rotates clockwise, the wood wraps around it. Below, the John C. Morris method, patented 1856, with metal band and end abutments to constrain the workpiece, and the stationary form to which both ends are bent while its middle remains clamped. In his patent reissue, Morris called Blanchard's method bending 'inward' and his own method bending 'outward'. (Drawing by R.B. Gordon.)

bill. In 1862 Blanchard had unsuccessfully petitioned Congress for a 14-year 'renewal' of this patent before turning to the option of a normal 7-year extension.[24] When younger, he had twice persuaded Congress he deserved 14-year renewals of his irregular turning lathe patent, in 1834 and 1848. The resulting 42-year duration of that patent was unprecedented and unrepeated.

To obtain the second renewal, Blanchard had resorted to somewhat deceptive non-verbal evidence in lobbying for passage of his bill: he obtained plaster busts of influential congressmen and used a sculpture-copying machine to make marble copies, which he exhibited in the Rotunda of the Capitol. He explained to the amazed and flattered congressmen that the copying machine was 'not a new invention, but a new application of an old one of mine, from which I have never realized much, and I want the patent renewed'. He apparently neglected to show them the machine, which was almost certainly very different from the ordinary Blanchard lathe whose patent they renewed. Rufus Choate quipped afterward that Blanchard had '"turned the heads" of Congress and gained his point' (Waters, 1881, p. 258). Perhaps the congressional branch of the invisible college provided a less effective learning situation (or more gullible pupils) than the others described above. Blanchard later obtained a French patent for his sculpture-copying machine, which resembled other sculpture-copying machines of its day in using a vertical-spindle cutter, not the horizontal-spindle cutter of the Blanchard lathe.[25]

In 1861 Congress lengthened the patent term to 17 years and abolished patent extensions by the Patent Office. This was in response to public agitation over perceived abuse of the patent system, especially by the Woodworth patent holders, who within three years in the mid-1840s obtained a reissue and two 7-year extensions (for a total of 28 years). Other patent law changes in 1861 and 1870 also practically eliminated reissues. Thus the invisible college of patent management proved itself capable, like other institutions of higher learning, of shutting down programmes deemed of dubious educational value!

CONCLUSION

Patent management was of course not the only social system within American society in which technical knowledge was generated during the nineteenth century. Other broad or narrow subsystems of society were also contributing to the dynamic of technological change; not least of these were the scattered but nonetheless highly important locations of commercial interaction collectively termed 'the marketplace.'[26] The government itself was another subsystem of society in which people gained technological expertise in choosing among technologies and developing new ones (Dupree, 1957; Smith, 1985). Of course the actual workshops, factories and mills where mechanics worked also functioned partly as schools in which they generated incremental innovation, or 'little kinks and devices' of technique.[27] Organizations such as the Franklin Institute in Philadelphia or the American Institute in New York City fostered interaction on questions of technology among inventors, mechanics and engineers, entrepreneurs, writers, educators and the public at large. Other social groupings that promoted growth of technical knowledge ranged from quite 'visible' schools and colleges where mathematics and sciences were taught (and only later in the century, engineering), through newspapers and magazines that reported technological developments, to ephemeral gatherings of people at industrial exhibitions (Curti, 1950; Post, 1983).

Such groups, of course, were not mutually exclusive; they could overlap.[28] Technology-related interactions taking place among people in such overlapping subsystems of society were not value-free; increasingly they were imbued with a burgeoning cultural value favouring technological 'progress'. They encouraged some individuals to attempt invention, and also helped them acquire a growing stock of technical knowledge and equipment with which to make a widening variety of inventions.

In a later era, when some businesses and some technological systems became very big and complex, and individual inventors were increasingly superseded by corporate laboratories, the comparative importance to invention of patent management and other relevant social subsystems probably changed in ways that are beyond discussion here. When land-grant and other universities started establishing engineering schools, the invisible college gained another structure on which to 'peg' or 'drape' itself.

Nevertheless, of the various subsystems that mediated between individual 'technologists' and the whole society in *ante-bellum* America, the patent management system was a very central one. 'Draped around' society's legal, economic and political systems, it had a relatively direct influence on the invention and development of particular technologies, and a special role in the development of technological cognition. Both detractors and supporters of the American patent management system have been at times vociferous ever since its inception 200 years ago. Some ascribe to it the very flowering of American technology; others say it has diverted effort and actually suppressed inventiveness. Its own participants may well have regarded it as an exasperating and costly way to proceed but, if we look past the surface to see how people actually interacted within this invisible college, we will find unexpected insights into the system's effects on the mobility of mechanics, the transmission of non-verbal information, and the social construction of invention.

Notes

1. For discussion of the relation of invention to patents from several perspectives, see articles in the special issue of *Technology and Culture* on patenting and inventing, October 1991.
2. Patent records include much more than statistics and specifications; for instance, the records of patent assignment and patent litigation are particularly rich in historical information. Some are noted in this chapter. For advice on retrieval and uses of different types of US patent records, see Reingold (1960) and Cooper (1987a).
3. The patent office reform of 1836 first established a staff of examiners to scrutinize applications for originality. From 1790 to 1793 the Secretary of State, the Attorney General and the Secretary of War (or any two of these) were supposed to decide whether a patent application described a truly original invention. No such examination took place from 1793 to 1836, when questions of a patented invention's originality was left to litigation. From one in 1836 and two in 1837, the number of patent examiner positions grew by 1861 to 12, and by 1870 to 22 principal examiners plus 44 assistant examiners. For useful brief accounts of these changes in the patent system, see Sherwood (1983) and Meier (1981).

4. For the minutes, see *Philosophical Transactions of the Royal Society of London*.
5. Derek de Solla Price (1975, pp. 102–3) suggested the invisible college members fell into conversation in the shops of makers of 'philosophical instruments' in London, as well as in taverns and coffee shops.
6. Members of 'IEG #1' exchanged memos through a central copying service, in 'only a couple of weeks instead of the several months of delay attending the formal publication, . . . at about $125 of subsidy per member per year' (Price and Beaver, 1966, p. 1012).
7. Crane (1972), p. 172, Figure 1. Crane thus links the life cycle of invisible colleges to the model of science propounded by Thomas Kuhn (1962).
8. The kind of 'writing' comprised in computer software is a modern exception to this generalization.
9. Interference suits took place before the granting of a patent, if there were two 'simultaneous' claimants to an invention; infringement suits took place when a patentee sued someone for using, making or selling a patented device without the patentee's permission.
10. The requirement of models ended in 1880. See Dood (1983, 1984, 1990). For the vicissitudes of making and housing models, see Ray and Ray (1974). The patent models are now scattered among museums and private collectors.
11. On Munn & Co. see Borut (1977).
12. As Dood (1990), p. 11, and others have pointed out, this 'learning by seeing' sometimes led to outright copying and fraud, before examination of patent applications for originality was instituted in 1836.
13. 'Patented' application files, containing evidence of this 'learning by trying', comprise part of National Archives Record Group 241.
14. Woodworth's assignment to congressman James Strong is recorded in *Digests of Patent Assignments*, vol. W-1, p. 4, National Archives Record Group 241. William W. Woodworth's papers, headed 'Receipts' and 'Account of expenditures and losses referred to in the annexed affidavit', are in *William W. Woodworth et al* v. *Hiram Casley et al*, US Circuit Court, Massachusetts District, October Term 1849, National Archives Boston Branch, Waltham, Mass., Record Group 21.
15. Assignees were outright owners of the intellectual property and could sell or rent it again, possibly subdivided, while licensees could use it but not resell it. Assignees could sue trespassers – called infringers – on their property; licensees lacked that right. A licence 'territory' could be very small (e.g., a 'shop right' for one machine shop). Licensees could not have their right sold out from under them, but the assignee to whom they owed royalties could change. Assignment rights were usually explicitly 'exclusive' and were gerrymandered around previous assignments.
16. The Blanchard Gunstock Turning Factory was incorporated 21 February, 1820; Blanchard assigned rights to the corporation for all purposes and locations on 29 June 1820. This was recorded in *Transfers of Patent Rights*, Liber F, p. 92, National Archives Record Group 241.
17. Collins Stevens described this 'learning by using' in his affidavit of 16 September 1856, in *Thomas Blanchard* v. *Warren Wadleigh and Isaac Lane*, US Circuit Court, New Hampshire District, 1857, Record Group 21, National Archives, Boston Branch.
18. Rosenberg (1982), p. 122, distinguishes between 'learning by doing', which takes place during production, and 'learning by using', which takes place during use of the product. In the case of machines that are both products and tools of production, the distinction gets hazy.

19. For an entertaining (fictionalized) description of the modern phenomenon among academic conference-going literary critics, see Lodge (1984).
20. William H. Seward was later President Lincoln's Secretary of State. Both he and Choate served terms in Congress as well. For entrée to evidence of this 'learning by suing' in pre-1860 infringement suits brought by Thomas Blanchard (irregular turning lathe), William W. Woodworth or his assignees, James G. and Jacob P. Wilson (planing machine), see reports in *Federal Cases*, where they are arranged alphabetically by plaintiff's name in volumes 3 and 30, respectively. Other Woodworth and Blanchard patent cases can be traced by use of *Shepard's United States Patent and Trademark Citations*. Original case files are held in various regional branches of the National Archives, Record Group 21.
21. For this episode of 'learning by suing' see Cooper (1991a), pp. 51–4. 'Froward' means perverse and obstinate.
22. It is also an example of the 'Matthew Effect' or the 'them that has, gets' tendency to give more credit to the already famous than to their lesser-known competitors. For discussion of this in science, see Merton (1988).
23. The testimony of witnesses Isaac Brock and Navy Lieutenant William Wood are in case files for *Jacob P. Wilson* v. *Daniel Barnum*, US Circuit Court Pennsylvania Eastern District, 1849, National Archives Philadelphia Branch, Record Group 21.
24. Papers of Senate patent committee, 37th Congress, Record Group 46, National Archives.
25. See 'Brevet d'invention de quinze ans, En date du 28 août 1855, Au sieur Blanchard, de Boston, Pour un machine à reproduire les bustes en marbre ou autres objets de sculpture avec réduction', in *Descriptions des machines et Procédés pour Lesquels des Brevets d'Invention ont été pris sous le Régime de la loi du 5 Juillet, 1844*, 2nd Series, vol. 51 (1855), p. 186.
26. For 'learning by selling' see Thomson (1987) and 1989).
27. For examples of 'learning by doing,' see Smith (1977), *passim*, esp. pp. 244–7, 283–6, and Malone (1988).
28. For attempts to use Venn or other diagrams to describe overlaps and hierarchies of social subsystems relevant to the construction of particular technologies, see Cowan (1989) and Constant (1989).

References

Borut, M. (1977) 'The *Scientific American* in Nineteenth Century America', unpublished PhD dissertation, New York University, pp. 89–151.

Bystryn, M.N. (1981) 'Variation in artistic circles', *The Sociological Quarterly*, 22 (Winter).

Chubin, D.E. (1983) *Sociology of Sciences: An Annotated Bibliography on Invisible Colleges, 1972–1981* (New York: Garland).

Constant, E.W. II (1989) 'The Social Locus of Technological Practice: Community, System, or Organization', in W.E. Bijker, T.P. Hughes and T. Pinch (eds), *The Social Construction of Technological Systems* (Cambridge, Mass.: MIT Press).

Cooper, C.C. (1987a) 'Thomas Blanchard's Woodworking Machines: Tracking 19th-Century Technological Diffusion', *IA, The Journal of the Society for Industrial Archeology*, 13.

—— (1987b) 'The Evolution of American Patent Management: The Blanchard Lathe as a Case Study', *Prologue, the Journal of the National Archives*, 19, 4 (Winter).

—— (1989) 'A Patent Transformation: Woodworking Mechanization in Philadelphia before 1856', paper presented 13 May, at 'Early American Technology – A Mid-Atlantic Perspective: A Conference in Honor of Brooke Hindle', American Philosophical Society, Philadelphia, Pennsylvania.

—— (1991a) *Shaping Invention: Thomas Blanchard's Machinery and Patent Management in Nineteenth-Century America* (New York: Columbia University Press).

—— (1991b) 'Social Construction of Invention through Patent Management: The Case of Thomas Blanchard's Woodworking Machinery', *Technology and Culture*, 32 (October).

Cowan, R.S. (1989) 'The Consumption Junction: A Proposal for Research Strategies in the Sociology of Technology', in W.E. Bijker, T.P. Hughes and T. Pinch (eds), *The Social Construction of Technological Systems* (Cambridge, Mass.: MIT Press).

Crane, D. (1972) *Invisible Colleges: Diffusion of Knowledge in Scientific Communities* (University of Chicago Press).

Curti, M. (1950) 'America at the World Fairs, 1851–1893', *American Historical Review*, 55.

Dood, K.J. (1983) 'Patent Models and the Patent Law 1790–1880', *Journal of the Patent Office Society*, 65, 4 and 5 (April and May).

—— (1984) 'Why Models?', in *American Enterprise: Nineteenth Century Patent Models* (New York City: Cooper-Hewitt Museum, Smithsonian Institution).

—— (1990) 'Patenting and Patent Models in Nineteenth-Century America', in Barbara Suit Janssen (ed.), *Icons of Invention: American Patent Models* (Washington, DC: Smithsonian Institution), pp. 11–13.

—— (1991) 'Pursuing the Essence of Inventions: Reissuing Patents in the 19th Century', in *Technology & Culture*, 32 (October).

DuBois, H.M. (1878) 'Bent Timber for Rims', *The Hub*, 20 (2 November).

Dupree, A.H. (1986) *Science in the Federal Government* (Baltimore, Md.: Johns Hopkins University Press; 1st edn, 1957).

Federico, P.J. (1960) 'Records of Eli Whitney's Cotton Gin Patent', *Technology and Culture*, 1 (Spring).

Ferguson, E.S. (1977) 'The Mind's Eye: Nonverbal Thought in Technology', *Science*, 197 (26 August).

—— (1980) *Oliver Evans, Inventive Genius of the American Industrial Revolution* (Greenville, Del.: Hagley Museum).

Johnson, W.A. trans. (1963) *Christopher Polhem, The Father of Swedish Technology* (Hartford, Conn.: Trinity College).

Journal of the Franklin Institute, vol. 3, first series (March 1829).

Kuhn, T. (1962) *The Structure of Scientific Revolutions* (University of Chicago Press).

Lodge, D. (1984) *Small World* (New York: Warner Books).

Malone, P.M. (1988) 'Little Kinks and Devices at Springfield Armory, 1892–1918', *IA, The Journal of the Society for Industrial Archeology*, 14, 1.

Meier, H.A. (1981) 'Thomas Jefferson and a Democratic Technology', in C.W. Pursell (ed.), *Technology in America* (Cambridge, Mass.: MIT Press).

Merton, R.K. (1988) 'The Matthew Effect in Science, II', *Isis*, 79 (December).

Mirsky, J. and Nevins, A. (1952) *The World of Eli Whitney* (New York: Macmillan).

Munn & Co. (1971) *The United States Patent Law* (New York).

Post, R.C. (1976) *Physics, Patents, and Politics: A Biography of Charles Grafton Page* (New York: Science History Publications).

—— (1983) 'Reflections of American Science and Technology at the New York Crystal Palace Exhibition of 1853', *Journal of American Studies*, 17.

Price, D.J. de S. (1954) 'Is Technology Historically Independent of Science? A Study in Statistical Historiography', *Technology and Culture*, 6 (Autumn).

—— (1971) 'Some Remarks on Elitism in Information and the Invisible College Phenomenon in Science', *Journal of the American Society for Information Science*, 22 (March–April).

—— (1975) *Science Since Babylon*, enlarged edition (New Haven, Conn.: Yale University Press)

—— and Beaver, D. deB. (1966) 'Collaboration in an Invisible College', *American Psychologist*, 21 (November).

Ray, W. and M. (1974) *The Art of Invention: Patent Models and Their Makers* (Princeton, NJ: The Pyne Press).

Reingold, N. (1960) 'U.S. Patent Office Records as Sources for the History of Invention and Technological Property', *Technology and Culture*, 1.

Robinson, E. and Musson, A.E. (1969) *James Watt and the Steam Revolution: A Documentary History* (London: Adams & Dart).

Rosenberg, N. (1982) *Inside the Black Box: Technology and Economics* (Cambridge University Press).

Sherwood, M. (1983) 'The Origins and Development of the American Patent System', *American Scientist*, 71 (Sept.–Oct.).

Sinclair, B. (1974) *Philadelphia's Philosopher Mechanics, A History of the Franklin Institute 1824–1865* (Baltimore, Md.: Johns Hopkins University Press).

Smith, M.R. (1977) *Harpers Ferry Armory an the New Technology* (Ithaca, NY: Cornell University Press).

—— (ed.) (1985) *Military Enterprise and Technological Change: Perspectives on the American Experience* (Cambridge, Mass.: MIT Press).

Thomson, R. (1987) 'Learning by Selling and Invention: The Case of the Sewing Machine', *Journal of Economic History*, 47 (June).

—— (1989) *The Path to Mechanized Shoe Production in the United States* (Chapel Hill, NC: University of North Carolina Press).

(Waters, Asa H). Anon. (1881) 'Thomas Blanchard, the Inventor', *Harper's New Monthly Magazine*, 63 (July).

Wilkinson, D. (1982) 'Reminiscences (1846)', in G. Kulik, R. Parks and T.Z. Penn (eds), *The New England Mill Village, 1790–1860* (Cambridge, Mass.: MIT Press), p. 88.

Part II
Diffusion, Learning and Ongoing Technological Change

The second part of this book focuses on the interactive relations among invention, diffusion and use. Citing examples from the history of industrial research, John K. Smith counterposes an evolutionary to a linear view of the history of technology. The linear view suggests a definite sequence, from invention to development to use, with the emphasis on the initating stage. Smith argues that a more circular view better explains technical change. The importance of a new technique depends on discovering a market niche, but the discovery of a niche often alters the direction and pace of future invention. Because market niches cannot be known when the invention process is beginning, the very purpose of an invention can only be fully known much later. The interactions around use that develop and select some techniques give technological change an evolutionary character.

Inventions can come into use in a variety of ways, and Ross Thomson considers how three kinds of use – within the shop of the inventor, via sale by single-product firms, and through sale by diversifying firms – distinguish economic forms of technological change. These forms are characterized by different learning processes and, as a result, different processes of technological change within industries, between industries and over time. Focusing on the activities of shoe manufacturing patentees, Thomson argues that opportunities for technological learning were highest for inventors associated with diversifying capital goods firms, lower for those gaining use with single-product capital goods firms, even lower for those using inventions in their own shops, and lowest for those without any kind of usage. Although communication occurs most readily within industries, much the same hierarchy exists between industries. Such interindustry communication allows us to understand technological change as an economy-wide process with discernible stages.

Social and technological dynamics which span economies may help to identify the conditions that lead productivity levels of countries to converge. Recognizing that some economic sectors are paramount in the generation of technological knowledge, Donald J. Harris constructs a model focusing on the growth of such a 'knowledge industry'. The output of this sector improves productivity in all sectors. Countries lagging in technological knowledge can advance by generating their own knowledge or by acquiring the knowledge of a leader. Technological convergence can occur if the follower builds a knowledge sector of adequate size to acquire or generate new techniques. As the knowledge gap closes, continued convergence will depend on a country's investment in and capabilities of knowledge improvement.

5 Thinking about Technological Change: Linear and Evolutionary Models

John K. Smith

For over three decades now historians of technology have been exploring the dynamics of technological change. Unravelling the paths by which new technologies are discovered and implemented has proved to be an intellectually engaging enterprise that has largely sustained the discipline. Much of the fascination with the process of technological change results from its complexity. On the other hand, this complexity has discouraged the development of more general approaches to the subject. Lacking explicit alternatives, most historians of technology have fallen back upon a linear model of technological change which links invention, development and commercialization in a sequential progression. Because the last two steps depend on the first one (invention), it is invention that has received the most scrutiny by historians. In fact, the concept of invention is tightly bound up with the linear model. Below it will be argued that the use of the linear model has distorted our understanding of technological change and that an alternative model, a Darwinian evolutionary one, more accurately expresses the complexity that emerges from the narrative histories.

THE LINEAR MODEL AND THE HISTORY OF TECHNOLOGY

The complexities of technological innovation have led historians of technology to emphasize *development* as the key to understanding technological change. Staudenmaier argues that this is a compromise between invention and innovation (Staudenmaier, 1985). One tradition in the history of technology focuses on the inventor as the key actor in the process of technological change. Economic historians, by contrast, tend to view technological change as a social process (or, as Nelson calls it, 'cultural evolution'). At the development stage, technological change contains both individual and 'environmental' aspects. A development team often includes the original inventor and is usually small enough for historians to be able to tell the story in terms of individual actors and actions. On the other hand, the goal of development is to create a technological artefact that will find a niche in the larger environment. Historians of technology have done a good job of showing how development gives new technologies their initial footholds outside the protected world of the laboratory.

The importance of development is stressed by Chandler (see Chapter 3), who argues that organizational capabilities make the process much less uncertain and unpredictable (Chandler, 1990). To translate an invention into an innovation requires skills in design, manufacturing and marketing. In a large organization these resources can be found in-house, thus reducing the cost and uncertainty of finding them outside. Nylon provides a good example of the role of organizational capabilities in innovation (Hounshell and Smith, 1988). In May 1934 chemists in Du Pont's central research department discovered a class of synthetic fibres that had outstanding properties. Over the next five years, the development process included selecting a particular fibre, working out processes for making it, and determining uses for it. To accomplish these goals Du Pont drew upon its expertise in organic synthesis, polymerization, chemical engineering and the fibre business (Du Pont was one of the largest manufacturers of rayon). The key person in the development process was their research director, Elmer K. Bolton, who defined his job as *innovation*; he kept one eye on the technological development and the other on the external environment in which nylon would have to find a place for itself. In 1940 nylon became a successful stocking fibre because Du Pont had the diverse skills necessary to take on a difficult technological challenge, knowledge about the fibre business, and an entrepreneurially-minded research director who brought everything together in a timely manner.

A shortcoming of this type of innovation story is that it ends when the product was *initially* commercialized. An implicit assumption in this approach is that from that point onwards further technological progress and new markets result from learning, which involves large numbers of small anonymous changes, and these do not generally interest historians of technology.

An attempt to develop a more comprehensive theory of ongoing technological change has been made by Hughes (Hughes, 1989). The central concept in his theory is the system which represents an interdependent network of technologies, and can include social and political components to sustain it. In this model, inventors create radically new technologies which will eventually evolve into systems. Other, more conservative, inventors work within the system by isolating specific problems and solving them systematically. One of the best examples of this latter type of invention was the transformer, which was critical to the success of alternating current electricity. Since the problem was so well defined, many inventors tried to solve it and several succeeded. Hughes calls this conservative invention, since the technological and economic objectives are so well defined, and these types of inventions do not lead to new systems. Another important aspect of Hughes's model is that the momentum that fully developed systems generate makes it extremely difficult for new radical inventions to evolve into parallel competing systems. Hughes's model is very helpful for thinking about technological change as ongoing evolution, but is teleological because we can only know which inventions are radical ones after they evolve into systems. From the historical perspective it is not usually possible to distinguish a radical from a conservative invention in its early existence. For example, probably one of the most pervasive systems in American history was the railway, which was ultimately outdone by the car and truck. Yet I suspect that, in the early phases of competition, the railways did not perceive the long-term threat that the internal combustion engine presented. In fact, the railways may have seen trucks as primarily feeders to the railheads, making trucks a complementary rather than competing technology. This example highlights an important aspect of technological change: that is, there continues to be a substantial degree of uncertainty in the process and its eventual outcomes. In recent decades Eastman Kodak did not appreciate the potential uses of photocopiers, and IBM also underestimated the appeal of personal computers. Any model of technological change, therefore, has to incorporate concepts such as learning, evolution and system, while not discounting novelty and unpredictability.

THE EVOLUTIONARY MODEL OF TECHNOLOGICAL CHANGE

A Darwinian model of evolution incorporates both these functions. It does so by disconnecting the generation of novelty (in nature) from the persistence of order (in nature). The connection between the two is the selection (by nature) of those individuals who have novelties that allow them to thrive in a particular environment (Basalla, 1989). As the anthropologist Kroeber pointed out, however, the analogy between biological and technological evolution is not exact because there is nothing comparable to a species in technology (Kroeber, 1923). All technologies can 'mate' and create offspring. Yet this distinction does not rule out viewing technological change as random variation and selection; cross-species 'mating' in technology serves to make the number of potential variations much larger and the process can occur very rapidly.

As a model for explaining technological change, the evolutionary approach avoids some problems that are inherent in the traditional 'linear' model. Historians of technology have become very adept at *describing* the complex set of circumstances that surrounded a particular innovation, yet to communicate their findings they tell the story in terms of a mostly linear process of invention, development and commercialization. This method of exposition gives the story an inherent structure, even if it is only an implicit one. Another bias that historians introduce into their work is that they usually study 'important' technologies: those which have become ubiquitous or perhaps notorious. Tracing the evolution of the technology back in time narrows the focus to an individual or a small group that gave this technology its initial push. In doing this there is a temptation to correlate the overall significance of the invention with the *eventual* significance that this technology has achieved. The pathway between invention and the current status of a technology is downgraded to an anonymous and largely automatic process of learning that takes on a degree of inevitability. In other words, invention becomes synonymous with innovation.

There have been cultural and ideological reasons for reducing the process of technological change to invention. It reinforces long-held American values that stress the importance of individual effort and enterprise. Our most celebrated inventor, Thomas Edison, created a Horatio Alger persona for himself that resonated with the cultural beliefs of his age (Millard, 1990). Not surprisingly he became the embodiment of the heroic inventor, even though his principal advantage over other inventors was his well-staffed laboratory which could quickly turn a concept into a new product, thereby beating potential competitors to the marketplace. Until recently historians have been more than willing to take Edison and other inventors (such as Eli Whitney) at their word. Thinking about technological change in terms other than heroic inventors and linear development requires a certain willingness not only to deal with complexity but also to go against cultural values.

The major difference between the linear model of invention, development and commercialization and the evolutionary model of variation and selection is the relationship between the various activities that in sum result in technological innovation. The linear model has an implicit hierarchy which not only puts invention first in the process, but makes it the most important step. In the evolutionary model there are two necessary steps, neither of which alone is sufficient for an innovation to occur. The first step is the generation of a technical novelty, and the second one is the selection into a niche in the external technological environment.

Interpreting technological change using these two models leads to different understanding about the process of technological change. For example, historians of technology frequently argue that more studies should be done of failed innovations (Staudenmaier, 1985). Failed innovations are important because they can be used to argue against the idea of technological determinism: that is, technological evolution is a *technological* process driven by the internal needs of the technology itself. The danger here, however, is that because the innovation failed there is a strong temptation to attribute that failure to technical factors. Electric vehicles are a case in point (Callon, 1987). To assert that they disappeared because of inadequate storage capacity of batteries is to say that the story can be understood without reference to any non-technical or external issues. In fact electric trucks were used in American cities until the 1930s. Historians generally have not taken up to the call for writing about failures because the complexity of a successful innovation disappears. It is this complexity that has made the study of technological change an intriguing discipline. With an unsuccessful innovation, all we have left is an artefact, the failure of which might be explained by any number of factors. Implicit in this approach is the idea that inventions should succeed – an implication that comes out of the linear model – and that failure must be explained. On the contrary, it is success that is unusual and requires explanation.

Another important difference from the linear model is that the evolutionary model separates the generation of technological novelties and their selection into the larger technological milieu, a process that may be called niche finding. Invention, then, is driven primarily by its own internal dynamic, at least partially disconnected from the larger world (Pacey, 1976; Basalla, 1989). Innovation occurs when certain technological novelties are *selected* from the wide array of choices and put into a niche in the larger technological environment. Thinking about technological change in this way leads to a number of important insights.

First is the proposition that necessity is not the mother of invention. The path from the definition of a problem to be solved to a successful innovation usually consists of an ongoing redefinition of both the *problem* and the *solution*, so that in the end the innovator ends up far from his initial goal. This result is not surprising when one considers the entire process of innovation. Choosing a problem on which to work involves surveying the existing world of artefacts and their *uses*. When the inventor decides to work on a particular technology, he or she must make assumptions about how the new technology will fit into the existing network of technologies and the social context in which they are used. In effect, in selecting a problem the inventor must remove a technology from its niche and develop a new artefact that he or she believes will also work in that niche. Importantly, many inventors fail to accomplish this direct substitution, and then must search for other potential uses for their creation. Ultimately success depends upon determining a proper niche for the new technology.

The innovation of celluloid plastics in the 1870s is an example of the dynamics of the process (Friedel, 1983). The Albany tinkerer, John Wesley Hyatt, apparently learned of a predicted ivory shortage, a material that was used in many applications including billiard balls. Hyatt set out to find a replacement for ivory billiard balls and began to experiment with soluble forms of cellulose, which chemists had discovered decades earlier. Technically, Hyatt discovered that the addition of camphor to nitrated cellulose gave the resulting material a mouldability and stability that earlier 'plastics' had lacked, yet this new material was so unlike ivory that billiard players rejected it. Instead of accepting failure, Hyatt re-evaluated his material and its properties. He found that his celluloid could be made to *look* like more valuable materials, such as ivory or tortoiseshell. This led him to sell his material, successfully, in a host of applications such as combs, brushes and mirrors. In the 1890s celluloid would find a new, less trivial, use as the film base for motion pictures. At the time there were no other candidates for the job. Without the subsequent selection of celluloid for use in motion pictures, Hyatt's celluloid might have been relegated to a lesser status for itself and its inventor.

There is an important exception to the usual ambiguity and complexity that an inventor sees as he surveys the world of technological artefacts and that is when the problem has already been defined in purely technological terms. These are the inventions that Hughes labels as conservative because the problem is so well defined within the systems context (Hughes, 1989). A good example is the transformer, the lack of which was holding back the widespread use of alternating current electricity. However, the acknowledgement of a technical problem does not ensure its solution. In this case, the components needed to make a transformer existed and a solution was rather quickly found by a number of inventors.

Another exception to the decoupling of invention and innovation is when an institution sponsors particular technical developments to achieve goals within a system. The military–industrial complex is one example of such a system. As David Noble has shown, the development of automated machine tools was sponsored by the military, and their use in industry began in the factories of defence contractors (Noble, 1989). Noble argues that this technology was selected because it gave management greater control over workers and engineers greater control over production processes. The economic benefits of this technology are uncertain and may be illusory. In the long run, automated machine tools can continue to be used in sectors where economic considerations are not critical, but in the world outside the military–industrial complex manufacturing is a highly competitive arena. Ultimately, to establish broader niches, these machines will have to prove themselves in the larger environment.

Following his system-based arguments, Noble has explicitly rejected an evolutionary model of technological change because it obscures the fact that modern technology is the product of a managerial elite that employs it to maintain and enhance its position of hegemony in America (Noble, 1989). In other words, all of America is a system. If this assertion is true, then evolutionary models do have limited explanatory power; however, under most other sets of assumptions which do not include a pervasive system, the evolutionary model is applicable.

Another reason for decoupling inventive activity from innovative success is that the creation of new technology is ultimately an act of creation (Pacey, 1976). The inventor must acquire technical inspiration from some tradition of technology, in much the same way that an artist belongs to a particular school. George Basalla argues that every artefact is directly descended from some antecedent: that is, every artefact is part of an evolutionary sequence (Basalla, 1989). If inventors, then, are working primarily within such a tradition, external influences such as economic demand are necessarily secondary. Cyril Stanley Smith and Arnold Pacey both argue that technological creativity springs from aesthetic and idealistic concerns (Pacey, 1976; C.S. Smith, 1976). These arguments favour a technology push explanation for innovation. Driven by aesthetic concerns, inventors create new artefacts, which may or may not find a niche for themselves outside the laboratory. Of course, an inventor can play both roles – creator and innovator – but, nevertheless, it is important to understand the different dynamic involved in the two processes. The major distinction between them is that in the laboratory an invention operates in a controlled environment in a function that the inventor has chosen, but once it becomes part of the larger technological matrix its usefulness and purpose is usually transformed through a process of ongoing selection and redefinition. Hyatt pushed celluloid out of the door as a replacement for ivory, tortoiseshell or even wood, but other inventors redefined it as a flexible film, a unique application (unlike earlier uses which could be easily substituted for by other materials). When Hyatt's celluloid became part of the inventor's vocabulary, it could be selected for other applications.

The evolutionary model provides concepts that are useful in understanding the dynamic of the uses of a technology once it has left the laboratory and become part of the technological matrix. In some ways this process is similar to that described by Hughes, who sees technologies becoming parts of large integrated systems. But technological change within his systems is predictable and orderly. Like biological evolution, technological evolution is orderly (that is, there are rules of selection) but there is an element of unpredictability that remains. Eventually the ongoing process of selection locates niches for a technology which prove stable in the long run. For example, the origins of aluminium lie in early nineteenth century attempts to isolate elements, which had been defined by Lavoisier in the 1780s (G.D. Smith, 1988). Because of the difficulty in reducing alumina ore to aluminium metal, it had a semi-precious status, 'silver from clay', until Charles Martin Hall discovered a cheaper refining process in 1886. Initially aluminium found markets as a decorative metal and in pots and pans. In either of these applications some other material could easily have replaced it. The popularity of aluminium for these uses derived somewhat from its novelty, which would not last long. (The American enthusiasm for novelty plays an important role in innovation by making the early tenuous niches more easy to develop. In the late 1950s, the hula hoop craze created a market for a new plastic, linear polyetheylene: see McMillan (1979), p. 90.) By the 1920s, however, aluminium became the material of choice for aircraft construction, in spite of its many shortcomings. By the Second World War new aluminium alloys had been developed that overcame most of the earlier problems. Today aluminium and aircraft are symbiotic technologies. This relationship creates a deep niche for aluminium. The word deep is used because it would be difficult to prise aluminium out of it. The design of aircraft is so bound up with the properties of aluminium that the introduction of an alternative material seems unlikely. Another important use for aluminium is the beverage can. This could be called a broad niche, because of the ubiquity of the artefact. At the same time, however, it could possibly be dislodged by some other material, notably steel or plastic. The stability of a broad niche depends upon economies of scale in production, making a technology the lowest cost alternative, and in the accompanying infrastructure of distribution and use. For example, a few years ago, the aluminium industry defeated efforts to reintroduce steel beverage cans by arguing that it would disrupt highly effective aluminium can recycling.

The concept of niche finding as ongoing evolution *separate* from the generation of new artefacts provides insight into the patent battles that have raged over most important inventions. Very successful *innovations* have occurred from finding a new niche for an existing technique. If this niche is a large and profitable one, several inventors usually surface, each claiming to be the original inventor. Although the patent laws only recognize technological novelty, sometimes judges have ruled against those who had actually made the thing at the earliest date, which represents an implicit recognition that niche finding can be more important than the generation of a novel technique. Many of the acrimonious conflicts between independent inventors and corporations can be reinterpreted as battles between initiators of new artefacts and those who continue their technological evolution – both in terms of the artefact itself and the ability to produce it on a large scale – and are adept at niche finding.

THE CASE OF INDUSTRIAL RESEARCH

The differences in the insights and interpretations that the linear and evolutionary models suggest are evident in the considerable body of scholarship that has been generated about the history of industrial research. Virtually all of this work has been done using a linear model of technological change. Before 1960 it was fashionable to view industrial research as systematic *invention*, replacing what had been an empirical, if not random, process during the heyday of heroic inventors. In this view scientifically-based invention was more efficient because theoretical understanding had replaced tedious searches for such things as a filament for an incandescent light bulb or a catalyst to convert nitrogen and hydrogen to ammonia (J.K. Smith, 1990). The shortcomings of this argument were explained by scholars in the late 1950s, who were able to show that numerous important industrial technologies had their origins in the work of independent inventors (Jewkes, 1958). One influential example of this scholarship was an article by Willard Mueller which asserted that Du Pont had invented only 10 of 25 important new products and processes that the company exploited between 1920 and 1950 (Mueller, 1962). His entire discussion was about invention, not innovation, and the inherent assumption he made was that invention was the necessary and sufficient component of innovation. In other words, the entire argument was based on the linear model of technological change.

The conclusion that industrial research laboratories were not invention factories set the stage for a new interpretation of industrial research, one in which the independent inventor and the corporation fought over the 'seeds' which, when properly planted would create prodigious wealth. Influenced by New Left scholarship, which asserted that big business has controlled American politics, society and technology, David Noble argued that corporations used their scientific and technological resources to harness technology for corporate ends (Noble, 1977). Rather than promoting technological change, corporations seek to control it to preserve the status quo and their existing investment.

One example of this phenomenon was the case of radio and AT&T. Leonard Reich argued that researchers in the AT&T laboratories delved into radio research out of fear that it might replace communication by wire (Reich, 1985). The goal was to control radio, not to promote it. In spite of AT&T's patents and lawyers, however, radio escaped to become an enormously successful technology in the 1920s. This happened largely because the real market for radios was broadcasting, which was discovered quite fortuitously in 1920, catching all the interested parties by surprise. In the end this story does not support the assertion that big business could contain innovation. From the evolutionary perspective this story is not surprising. The process of niche finding is inherently uncontrollable and will always contain a certain degree of unpredictability.

Thomas P. Hughes has developed a dynamic model to explain the relationship between inventors and corporations (Hughes, 1989). Independent inventors produce radical inventions that lead to the creation of large-scale technological systems which, over time, gain technological momentum and represent a large amount of invested capital. Because of these latter factors, additional technological change is conservative in that it improves the existing system rather than creating a new competing one. Eventually, these technological systems add socio-political superstructures to protect themselves and make the prospects for competing systems even smaller. One implication of this model is that at some point the pervasiveness of system is so great that radical invention becomes impossible. Hughes perceptively argues that the critics of technology in the 1960s, such as Jacques Ellul, believed that large technological systems dominated the industrial world, and that the survival of these systems was the foremost value in industrial societies (Hughes, 1989, Ch. 9).

Hughes's model is a linear one because his definition of a radical invention is one that leads to a system; the two are linked by a historical process. There is not much room for unpredictability in Hughes's world of order, system and control. Like the New Left historians, Hughes realizes that this is what capitalist businessmen or system-building entrepreneurs wanted, but there is a large gap between what they desired and what they were actually able to achieve. From the perspective of the evolutionary model, the process of innovation can never be totally controlled. Although Hughes's model works for electrical power systems, it is not so successful for another massive system, the railway, which was undermined by cars, trucks and eventually aeroplanes.

In regard to industrial research, Hughes begins his chapter with the tragic story of the battles between Edwin Armstrong and RCA that contributed to his suicide. The moral of the story about the relationship between independent inventors and corporations is clear. Yet, at the end of this chapter, Hughes notes that independent inventors are responsible for 'air conditioning, automatic transmissions for automobiles, power steering, catalytic cracking of petroleum, cellophane, the jet engine, Kodachrome film, magnetic recording, Polaroid Land cameras, quick-freezing and Xerography' (Hughes, 1989, p. 183). From this evidence, then it appears that existing systems have not been able to stop a number of important innovations. This apparent contradiction between a corporation's desire to control the direction of technological change and its frequent failure to do so can be best explained by the evolutionary model, which recognizes the inherent unpredictability of the process of technological change.

One of the most famous, important and investigated innovations was the vacuum tube, the key technology for the development of radio. It is possible to see the story as independent inventors (John Ambrose Fleming, Lee de Forest and Edwin Armstrong) supplying large organizations with this essential device. Since de Forest called himself 'the father of radio' and felt cheated by AT&T because of terms under which he sold the rights to his device, this episode can be interpreted as a conflict between an independent inventor and a predatory corporation. Instead of interpreting this story using the linear model of technological change, however, the evolutionary approach leads to different conclusions. Hugh Aitken's painstaking reconstruction of the events surrounding the development of the vacuum tube reveals an evolutionary process (Aitken, 1985, Ch. 4).

Technologically the vacuum tube was connected to the earlier development of the light bulb. The basic phenomenon, the flow of current from a filament across a vacuum, was first observed by Edison. In part because the electron had not been discovered, Edison did nothing with this observation. Next, Fleming in Great Britain utilized the directional character of the flow of electricity from the filament to produce a two-element diode. A few years later Lee de Forest added a third element which could be used to regulate the flow of current from the filament to a positively charged surface. Years later, after the importance of these devices had been established, de Forest, Fleming and Armstrong engaged in monumental patent battles in which millions of dollars in royalties were at stake for the inventors and the companies that had acquired the rights to their inventions.

For de Forest it was critical to prove his 'audion' was not just an improvement of Fleming's 'valve'. On this issue Aitken makes an important point. Although the two devices may have been physically similar, he argues that each man *understood* his device differently, which led each inventor to very different perceptions of its utility. De Forest had been searching for a device that would amplify radio signals, and his audion, *when reworked using new concepts*, later proved to be capable of performing that function. But de Forest did not take that step because he understood his device in terms of ionized gases, not in terms of emission of electrons. His conception of the audion led him to believe that ionized gas molecules were responsible for the current, so he did not fully evacuate the bulb.

The success of the audion as an amplifier of radio waves was the result of the electronic intrepretation of its operation by Irving Langmuir at GE and H.D. Arnold at AT&T. Their new insights prompted them to redesign the audion and remove the gas from it. According to Aitken, this was not just the *development* of de Forest's audion, it was the invention of the vacuum tube. As Aitken points out, 'Later generations of vacuum tubes traced their ancestry to the "hard" vacuum tubes developed by the laboratories of General Electric and the telephone company. But these were already an important evolutionary step beyond de Forest's audion' (Aitken, 1985, p. 201). It is ironic that de Forest actually developed the right device for the right job, but for the wrong reasons. Consequently, he fell short of being the father of radio. Not only did the vacuum prove to be the ideal radio receiver, but also, as Armstrong discovered, it was essential for broadcasting. Finally, the vacuum tube was what AT&T needed to make long-distance telephony practical. In sum, de Forest's device was part of a technological sequence of artefacts and was ultimately 'selected' for applications other than those that he choose.

The difference between technological novelty and successful innovations became a critical issue for Du Pont's Duco lacquers, which revolutionized car finishing after 1924 (Hounshell and Smith, 1988, Ch. 6). Duco could be quickly applied and made durable colour finishes possible. There was virtually nothing that was technically new in what Du Pont had done and, when the company sued infringers of its patent, the defendants had no trouble finding numerous anticipations of Duco. After losing the patent in initial hearings, Du Pont had it restored when the judge dismissed the technical argument and ruled that a patent monopoly could be justified on the basis of finding a new use for an existing technology. In this he recognized that the creative step in an innovation can be in finding a market niche. The actual invention of Duco-type lacquers meant nothing until a Du Pont chemist discovered that they made a good car paint.

Another important Du Pont example (one that is cited by both Mueller and Hughes as an outside invention) is cellophane, the most profitable non-fibre product in the company's history (Hounshell and Smith, 1988, Ch. 8). The traditional inventor of cellophane was Jacques Brandenberger, who made a clear regenerated cellulose film using the viscose process, which had previously been used to make rayon fibres. Brandenberger's contribution to this technology was to develop machinery to extrude thin films, a feat which required considerable mechanical ingenuity. Du Pont purchased the American rights to this technology in 1923, when a small amount of cellophane was being imported as a decorative overwrap for items such as the famous Whitman Sampler. Initially, markets for the material turned out to be disappointingly small because cellophane had an important technical deficiency; it allowed the free passage of water vapour, making it useless as a food packaging material. After learning of this problem, Du Pont launched a crash programme to develop a moisture-proof coating for cellophane. The technical requirements for this coating were quite exacting. After a young Du Pont chemist, Hale Charch, solved this problem, cellophane sales and profits took off.

This success could not be completely attributed to this innovation and the skill of Du Pont's marketers. The idea that packages were advertising and not merely containers had gained momentum during the 1920s, but Du Pont could not have foreseen that the Great Depression would lead to the replacement of the corner grocer with the large self-service supermarket. The supermarket concept was developed by King Cullen, who opened his first store in 1931. In this new type of shop, instead of having an assistant retrieve goods from high shelves and make suggestions about the relative merit of various brands, the shopper was now on his or her own. Cellophane found increasing use in attractive packaging that sold all kinds of goods, but especially impulse items such as cakes, sweets and nuts. In contrast, in Brandenberger's France, where the supermarket did not replace small speciality shops, the use of cellophane remained small. Du Pont's phenomenal success with cellophane ultimately depended upon the ability of research management to match this technology to an important change in food marketing.

Du Pont's success with cellophane had been overshadowed by that of nylon, which has become a paradigmatic science-based innovation (Hounshell and Smith, 1988, Ch. 12), but even it developed along an evolutionary and non-linear path. The research that led to nylon began with a purely scientific exploration of polymers, with the expectation that the knowledge gained would be used to improve the company's polymer-based products, which included Duco, rayon and cellophane. The research of chemist Wallace Carothers accomplished this objective when he developed a method for preparing polymers in an unambiguous and controlled manner. The course of this work dramatically changed direction in April 1930 when two of Carothers's associates unexpectedly discovered a rubber-like polymer and a synthetic fibre. The path these discoveries took to become the successful products neoprene and nylon contained many twists and turns. Technically, the commercial production of both was a *tour de force* of chemistry and engineering, but nylon has earned profits of a different order of magnitude from neoprene. The latter began and stayed a speciality rubber, and has had to compete with an increasing number of similar materials. Nylon, however, has been used in many different applications. Its first important market was in what has become known as nylons (stockings), a significant but limited market in 1940 when they were introduced to the general public. After the Second World War nylon found uses in other apparel applications, tyre cord and as a structural plastic (Hounshell and Smith, 1988, Ch. 19). In the 1960s the development of bulked, continuous filament caused demand for nylon to skyrocket in the ensuing carpeting of America. This new carpet-making technology, which was developed outside Du Pont, was responsible for much of the profit that nylon has earned. Although nylon was a scientifically and technologically important product from the start, it was the ongoing development of the material and new niches for it that has made it the most important product in Du Pont's history.

Nylon was all the more remarkable because the managers in charge of its development selected an appropriate target niche, silk stockings, rather early. On the other hand, this is not too surprising since in the 1930s rayon had attacked silk in virtually all of its applications except stockings. Rayon did not make good stockings because of a lack of elasticity, especially when moist. In contrast to nylon, Du Pont's other fibres experienced a longer and less straightforward selection process.

Du Pont's acrylic fibre, Orlon, technically depended on the discovery of solvents for the previously intractable acrylonitrile polymer (Hounshell and Smith, 1988, Ch. 19). The initial evaluation of the fibre showed that it had good resistance to weathering but that it could not be easily dyed. Based on these properties, Du Pont marketed the fibre for outdoor uses, such as awnings. The textile industry, however, discovered that Orlon could be spun into a wool-like yarn, the application that eventually became the primary market for the fibre, after Du Pont worked long and hard to solve the dyeing problem.

A final example is polyester, which has had a complex technical and use history (Hounshell and Smith, 1988, Ch. 19). The story begins in the mid-1930s when Wallace Carothers asked one of his assistants to polymerize terephthalic acid to produce a polyester that would be resistant to water. The chemical reasoning behind this experiment was straightforward, but problems occurred in the polymerization. Because of the large amount of work needed for nylon, Du Pont chemists never got around to trying this experiment again. In 1940 two British chemists working for the Calico Printers Association read Carothers's paper that reported the failed experiment. They then repeated it and produced a fibre with outstanding properties. Not knowing what to do with it, they turned it over to ICI, which soon gave Du Pont rights to produce it in the USA. After the war Du Pont worked out the chemistry to make polyester and discovered that, through various physical treatments, fibres with a wide range of properties could be produced. The big question that du Pont faced was which properties could be profitably exploited? Initially, Dacron polyester was targeted as a wool substitute, largely because that had been a long-time research goal. The new fibre had only limited success in this application. Eventually, polyester had to go through several stages, including the ill-fated double knits, before a satisfactory niche was established. The phenomenonal success of the fibre resulted from a fibre customer who discovered polyester's wrinkle resistance, which led to blended fabrics that have the feel of cotton and wool, but with superior wrinkle resistance. The incredible versatility of polyester fibres has opened up many applications for it. Recently, very fine polyester fibres that have a silk-like look and feel have been introduced in Japan and Europe (*New York Times*, 15 March 1991). Overall, the complexity of the polyester story cannot be adequately conveyed in a simple linear model of invention, development and commercialization. The more flexible evolutionary model is consistent, however, with ongoing technological change and market selection.

When firms such as Du Pont have used the linear model themselves, it has led to problems. For example, Du Pont came to interpret the success of nylon as emanating from the sheer novelty of the product. In other words, innovative success was equated to the degree of novelty of the product. The flaws in this approach became evident in Du Pont's handling of the super-fibre, Kevlar, which is used for such things as bulletproof vests, military helmets and lightweight canoes (Hounshell and Smith, 1988, Ch. 25). From a technical standpoint, Kevlar is a remarkable product. However, from a market or use perspective it has been a major disappointment: Kevler has yet to develop markets large enough to pay off Du Pont's enormous R&D expenditures and plant investment. Because of its technical uniqueness, Kevlar will probably find significant niches in the future, but by then Du Pont's accumulated losses on the product will be very large.

CONCLUSION

It is the ongoing process of technological adaptation and market niche recognition that are the strengths of industrial research. Change and complexity define the nature of twentieth-century technology and economy. The firm represents a microcosm of the larger world and contains knowledge about the generation of technological novelties (through research) and the configuration of the larger technical and economic world (through diversified product lines). Successful innovation occurs when a company can produce a new product that finds a significant niche in the larger technological environment. The degree of technical novelty in that new product is not directly related to the ultimate 'usefulness' of that technology. In other words, an important innovation can occur from finding new uses for older things. The degree of novelty is important in a company's ability to develop a proprietary advantage over its competitors. The American patent system perhaps has distorted the innovation process by recognizing and rewarding technical novelty at the expense of those who are adept at finding niches for existing technologies or combinations of existing technologies. The philosophy of the patent system is an important example of how the linear model of innovation has had an important impact on the process of innovation itself. I believe that the evolutionary model avoids the implicit simplification of what is an inherently complex and unpredictable process.

References

Aitken, Hugh G. (1985) *The Continuous Wave: Technology and American Radio, 1900–1932* (Princeton, NJ: Princeton University Press).

Basalla, George (1989) *The Evolution of Technology* (New York: Cambridge University Press).

Callon, Michel (1987) 'Society in the Making: The Study of Technology as a Tool for Social Analysis', in Wiebe E. Bijker, Thomas P. Hughes and Trevor Pinch (eds), *The Social Construction of Technological Systems: New Directions in the Sociology and History of Technology* (Cambridge, Mass.: MIT Press).

Chandler, Alfred D., Jr (1990) *Scale and Scope: The Dynamics of Industrial Capitalism* (Cambridge, Mass.: Belknap Press).

Friedel, Robert (1983) *Pioneer Plastic: The Making and Selling of Celluloid* (Madison, Wis.: University of Wisconsin Press).

—— (1986) 'Silver From Clay', *American Heritage of Invention and Technology*, 1, 3 (Spring).

Hounshell, David A. and Smith, John Kenly, Jr (1988) *Science and Corporate Strategy, Du Pont R&D 1902–1980* (New York: Cambridge University Press).

Hughes, Thomas P. (1989) *American Genesis: A Century of Invention and Technological Enthusiasm* (New York: Viking).

Jewkes, John, Sawyers, David and Sillerman, Richard (1958) *The Sources of Invention* (New York: St Martin's Press).

Kroeber, Alfred L. (1923) *Anthropology* (New York: Harcourt Brace).

McMillan, Frank M. (1979) *The Chain Straighteners* (London: Macmillan).

Millard, Andre (1990) *Edison and the Business of Innovation* (Baltimore, Md.: Johns Hopkins University Press).

Mueller, Willard (1962) 'The Origins of the Basic Inventions Underlying Du Pont's Major Product and Process Innovations, 1920–1950', in *The Rate and Direction of Inventive Activity: Economic and Social Factors* (Princeton: Princeton University Press).

Noble, David F. (1977) *America By Design* (Oxford University Press).

—— (1989) 'Automation Madness or the Unautomatic History of Automation', in Steven L. Goldman (ed.), *Science, Technology, and Social Progress, Research in Technology Studies*, vol. 2 (Bethlehem, Pa: Lehigh University Press).

Pacey, Arnold (1976) *The Maze of Ingenuity: Ideas and Idealism in the Development of Technology* (Cambridge, Mass.: MIT Press).

Reich, Leonard S. (1985) *The Making of American Industrial Research: Science and Business at GE and Bell, 1876–1926* (New York: Cambridge University Press).

Smith, Cyril Stanley (1976) 'On Art, Invention, and Technology', *Technology Review*, 78 (June).

Smith, George David (1988) *From Monopoly to Competition: The Transformation of Alcoa* (New York: Cambridge University Press).

Smith, John Kenly (1990) 'The Scientific Tradition in American Industrial Research', *Technology and Culture*, 31, 1.

Staudenmaier, John M. (1985) *Technology's Storytellers: Reweaving the Human Fabric* (Cambridge, Mass.: MIT Press).

6 Economic Forms of Technological Change

Ross Thomson

Institutions generating new techniques form a principal engine of capitalist growth, but what determines the output of this engine? One promising answer, with a heritage extending back to Adam Smith, interprets technological change as an outcome of the very growth processes it engenders. In a positive feedback system, economic growth leads to expanded markets and learning which fuel further technological change.

Whether such fuel can power ongoing invention and growth, however, depends on the design of the engine: that is, on the organization of those who generate and diffuse new techniques. Moreover, design improvements can augment the efficiency of the engine and thus strengthen the growth-invention feedbacks.

This chapter will identify how economic institutions could form such a feedback system and argue that they did so in the case of the US shoe industry in the second half of the nineteenth century. Three economic forms organizing technological change are identified, and it will be argued that these forms structured technological learning and innovation in three increasingly wide-ranging ways. First, the economic forms of technological change each differently influenced the pace, direction and cumulativity of technological change within industries. Second, they varied in the manner and the extent to which they related technological changes in one industry to those in the rest of the economy. Finally, the forms periodized technological change within industries and in the economy as a whole.

FORMS, BARRIERS AND PROCESSES

The most important shoemaking invention – the McKay sole-sewing machine, which stitched together the uppers and soles of shoes – illustrates each of the three forms. Its inventor, Lyman Blake, initially intended it for use in his shoe factory, and for two years it was used and developed there. This is the form of *self-usage*, where techniques are generated within the firms that intend to produce with them. But then Blake sold the patent to Gordon McKay, who started a firm to develop and sell the machine. This is the form of *commodity usage*, in which the invention is embodied in and diffuses as a capital good made by a firm specializing in its production and sale. McKay's success with the sole-sewing machine provided the means to diversify into other shoe machines for lasting, nailing and heeling. This is the form of *diversification*, where established firms use technical change to expand their product lines. As McKay's sole-sewing machine shows, economic forms of technological change are distinguished by the kind of economic institutions generating them and by the attendant modes of diffusion.[1]

Prospects

The meaning and significance of these economic forms are best seen by relating them to the obstacles that any technical change must overcome. Prospects for technical change vary with the capacity and need to solve technical problems *and* with the communication of knowledge among those who generate, diffuse and introduce new techniques (Thomson, 1986, 1989b). Any technical change applies skills and equipment to perceived needs, but to interpret technical change as a direct response of capacity to need would fundamentally misunderstand both the obstacles to technical change and the process of overcoming them. The necessary financing, need identification, inventive capability, production skills and marketing abilities are virtually never united in one person, and neither are they easy to acquire from others. Socio-economic barriers to communication can thus block technical change. Then the critical question for understanding technological change is whether social mechanisms exist that can overcome communication barriers, barriers arising from the very specialization that Adam Smith associated with productivity growth in market economies.

Economic forms vary in how and how well they overcome barriers to successful invention. Since the firm inventing for self-usage is involved in production, it clearly perceives the technical problem to be overcome. It faces no marketing difficulties because it will introduce the technique only into its own production process (although marketing problems may arise if the process alters its product). But the remaining barriers can be debilitating. The firm may be too small to meet costs of development, to build large enough plants, or to sell enough to justify the innovation. The technical knowledge required to solve a problem is quite different from that needed to identify one. The firm may not employ workers with such knowledge or normally contract with firms that could supply this expertise. Furthermore, monopolization efforts by firms and obstacles to worker mobility may limit the diffusion of new techniques.[2]

Communications barriers to technological change often arise from the separation between industries. One industry can hold knowledge of techniques needed in another industry but have no way to learn of the other's need. Because they are undertaken within a single industry, inventions intended for self-usage face particularly difficult obstacles to interindustry communication.

Commodity usage overcomes some of these barriers but can exacerbate others. Innovating capital goods firms possess knowledge needed to solve problems that potential users face, although their separation from users makes the identification of problems a new hurdle until regular marketing channels exist. Knowledge acquired by former managers and workers of firms using similar technology to equip plants for other industries can mitigate the interindustry communications difficulties plaguing self-usage. Finance and marketing may be more difficult because new firms cannot call on past profits and must form marketing systems. However, commodity usage greatly facilitates diffusion. Whereas the self-using firm commonly tries to limit diffusion, the commodity-generating firm tries to spread the new technique throughout the pool of potential users; its incentive to invent is not limited by the plant capacity of any one user. Commodity usage has particular advantages when inventive and production skills vary from those of using industries, when users are small relative to development costs, and when users are numerous enough so that none individually can justify development costs.

Diversification can transcend limits faced by both other forms. Unlike new firms, diversifying firms can apply already-acquired profits, technical skills, production facilities and marketing systems to the new product. Their sales efforts can help identify new technical problems. They may already learn from other kinds of capital goods producers, although information still flows to them most readily from users and competitors.

Processes

The three forms of technological change influence not only the prospects for success but also the process through which success is pursued. Each form contextualizes technological change and situates it in relation to what precedes, follows and surrounds it. An invention thus is a part of an ongoing process that ties past to future, inventor to inventor and firm to firm.

The institutional forms that condition innovators also put them in contact with others in similar conditions. As it proceeds, technical change characteristically increases the number and changes the kinds of involved firms and inventors. Early attempts at diffusion are a particularly important means of spreading learning to others. Successful invention is hardly a matter of isolated perseverance.

Since forms of technical change vary in their method of diffusion, they spread inventive activity differently. Self-usage is the most circumscribed. It spreads knowledge of the invention within the firm and perhaps to a few contractors outside. Those who know of the process may take it to other firms or form their own companies. An innovation that discernibly changes the product spreads awareness more widely but not necessarily with knowledge of how to duplicate it. Particularly when there are many firms, secretive innovators can thwart not only diffusion but also the learning needed to bring techniques to practicality.

Commodity usage has the opposite goal: to make the technique (embodied in a new commodity) known as widely as possible through sales. The demonstrations, training, complaints, servicing and promotion that attend sales spread knowledge of the new technique to potential users and to potential inventors. On the one hand, the diffusion of technological knowledge fosters invention by new inventors and new firms. Some new inventors are drawn from other industries, which increases the likelihood that technological affinities among industries will direct technological change. On the other hand, such learning deepens the knowledge and sustains the inventive activity of established firms and inventors. Through this market-mediated process, which I have called *learning by selling*, technological change becomes cumulative (Thomson, 1987, 1989b, esp. pp. 80–2, 94–6, 235–42).

Diversifying firms combine learning by selling with their own greater resources. Their profits, skilled employees and marketing systems enable them to develop techniques within the firm. In this they resemble self-users. Marketing systems allow them to sell their products more quickly and widely, which in turn supports further invention by them and others. Diversifying firms are well situated to develop the techniques others have invented, so they often purchase patents or whole firms with this goal (Levine, 1981, Ch. 4; Shapiro, 1986).

Through these forms, technical change follows its own cumulative process. Whatever form is taken, useful improvements typically involve a series of problems, inadequate solutions and a learning process that continually reformulates problems and arrives at new solutions. The success of this exploration depends on its socio-economic organization. To develop and diffuse the invention, an inventor interacts with others, some of whom later invent; the number of inventors and firms thus grows. Practicality, perhaps together with marketing and managerial innovations, can lead to the firm's growth and ongoing invention. Inventions used as capital goods have a particularly pronounced cumulative impetus. Not only does learning by selling spread knowledge and invention, but networks in the capital goods sector and in the community of inventors formed around that sector communicate technical problems and solutions between occupations and industries (Thomson, 1991).

Changes in the numbers and sizes of firms are supplemented by qualitative changes in the form of technological change. As development proceeds, self-usage may give way to commodity usage or to diversification, and commodity usage can bring about more invention for self-usage as well as the entry of diversifying firms. Both the expansion of inventive effort and the changing forms of that effort structure the path of technological change.

Such cumulative processes give technical change an autonomy from the broader economy. Ongoing technical change might depend more on earlier innovations along the same path – marketing as well as technological – than on changes in needs and capacities in the outer economy. Moreover, forms of technological change cannot be inferred from the state of social needs and capacities. Forms are not simply arrayed in front of an inventor, who chooses the one that maximizes potential productivity advance. Institutional innovations are made in order to overcome problems with older forms, not to realize some 'optimal' form. Economic forms play an irreducible role in the process of technological change.

THE CASE OF US SHOE MANUFACTURING

One feature of the McKay's experience can be generalized to all shoemaking: all three forms contributed to technical change. Self-usage was prominent in the craft period and remained common throughout the country. Shoe firms initiated the division of labour, improved craft skills, introduced the pegged shoe and devised machines and tools for their own use. The movement of shoe producers spread technical and managerial innovations within and between localities. Shoe firms also improved the design and utilization of purchased machines.

Commodity usage was the predominant form through which shoe machines originated. Ever since Elias Howe, inventors intended the sewing machine for sale, and by the 1870s hundreds of thousands of these machines were sold annually. Other machines also followed the path of commodity usage, including waxed-thread sewing machines for stitching uppers, pegging machines, McKay and then Goodyear bottom-sewing machines, and dozens of other machines to mould, cut, last, nail, burnish and polish.

Diversification arose out of the path of commodity usage. Diversifying firms, including McKay, used profits and innovative experience from earlier commodities to develop new products that supported future growth. Major sewing machine companies (such as Singer, Wheeler & Wilson and Willcox & Gibbs) followed this path by developing specialized machines to stitch leather, hats, buttonholes, carpets and books. Shoe machinery firms followed suit, including Elmer Townsend's companies, which moved from waxed-thread sewing machines to pegging, eyeletting and trimming machinery. Other firms diversified to overcome limits to their machines posed by complementary operations. Most importantly, the Goodyear Boot and Shoe Sewing Machine Company, which made the second major sole-stitching machine, developed a whole system of bottoming machinery to work with its sole stitchers.

The Hierarchy of Economic Forms

Inventors following the three forms faced different barriers, developed their inventions in distinct ways, and accordingly varied in their further inventive activity. The logic of the difference has already been suggested: in an industry of many firms with narrow technological experience, self-usage limits access to knowledge, provision of finance and diffusion of techniques. Commodity usage overcomes some of these limits but remains constrained by new firms' lack of funds and knowledge. Diversification reduces these obstacles. Thus, in their capacity to sustain a process of technological generation, a hierarchy of forms exists: diversification is the most fruitful, commodity usage is next, and self-usage is the least fruitful.

Patent data provide evidence to evaluate the existence of this hierarchy. A sample of 264 inventors with US shoe manufacturing patents from 1848 through 1901 was classified by the kind of usage they gained for these patents. These inventors, all US residents, comprised one-tenth of all shoe

Table 6.1 Shoe manufacturing patenting by type of use, 1848–1901 (numbers of patents)

	Self-used	*Commodity usage*	*Diversification*	*Unused*
1 Total patentees	44	64	37	119
2 Multiple patentees	21	46	31	33
3 Repeat patents	36	262	394	65
4 Frenquency (2/1)	0.48	0.72	0.84	0.28
5 Extent (3/2)	1.71	5.70	12.71	1.97
6 Rate (3/1)	0.82	4.09	10.65	0.55
7 Share with both shoe sewing and shoemaking	0.02	0.11	0.35	0.04

Notes: Multiple patentees were those with more than one US shoe manufacturing patent. Repeat patents are those shoe manufacturing patents after the inventor's first. Use types are defined in the text. Frequency refers to the share of inventors receiving repeat patents. Extent and rate are ratios of repeat patents to multiple patentees and all patentees, respectively.

Sources: Patents were first identified from US Department of Commerce, *U.S. Patent Classification – Subclass Listing*, class 12 and leather sewing machines from class 112. Using annual reports of the US Commissioner of Patents, additional shoe manufacturing patents were located. Use was determined from patent grants and city and business directories for 80 US cities.

manufacturing patentees in this period. (Later in this chapter 73 other self-using patentees are added.) Inventors who owned or managed diversifying firms or assigned one or more patents to these firms are classified as diversifying. These firms included eight major companies that diversified widely and another eight that spread among two or more types of shoe machines, tools or other shoemaking inputs. Commodity usage includes inventors who owned, managed or assigned to non-diversifying capital goods firms. Self-usage includes those whose inventions were not sold as capital goods but could have been used within innovating shoe factories because the patentee or assignee was a shoe factor owner, superintendent or foreman at or after the time of the patent. Finally, patentees that fell outside these three usage categories were considered to have not gained use.[3] Inventors followed all three forms of technological change: 44 gained self-usage, 64 commodity usage and 37 diversification (see Table 6.1).

The experience of shoe patentees shows that differences in ongoing invention were associated with, and thus may have been occasioned by, different kinds of usage. The hierarchy is clearest for the frequency of repeat patenting (the share of patentees taking out more than one shoemaking patent). The share of inventors with repeat patents declines from 84 per cent for diversifying inventors to 72 per cent for those with commodity usage, 48 per cent with self-usage and 28 per cent without usage.

Furthermore, among those continuing to invent, the form of usage powerfully influenced their number of repeat patents: diversifying inventors had twice as many as those with commodity usage and seven times the number of those with self-usage and without use (see line 5). This difference was independent of the occupations of inventors prior to shoemaking patenting; within usage categories, shoemakers had similar rates of repeat patenting as had machinists and other inventors.[4]

The superiority of diversification as a form is also manifested by the greater patenting breadth of diversifying inventors. Over one-third of diversifying inventors moved between shoe-sewing and other shoemaking patents, compared to 11 per cent for inventors with commodity usage and 2 per cent for those with self-usage (see line 7).

Complementarity and Cumulativity

Separately and together, these forms made technical change ongoing, as the McKay machine exemplifies. In the mid-1850s, Lyman Blake was a foreman in the shoe factory in which he hoped to use his machine. He was already dependent on purchased machinery; his company purchased Singer dry-thread and Townsend waxed-thread sewing machines, and he had been trained at a Singer agency. Learning by selling thus supplied knowledge needed for Blake's invention (Thomson, 1989b, pp. 156–65, 214–17).

Soon after introducing his bottom-sewing machine in his company's factory, Blake found that he did not have the funds, the knowledge or the production skills to bring his machine to practicality. So he sold the patent to Gordon McKay, a well-connected Boston machinist and printing press inventor and producer. This sale was an important instance of how efforts to gain usage spread information across industry lines. McKay sought to develop the machine through his own inventive efforts and those of Blake and hired machinists. The ties to self-usage continued. Blake worked on the machine in a South Shore shoe factory and, along with three prominent Lynn manufacturers, tested it on women's shoes. Most importantly, during the Civil War McKay and Blake themselves used their machine to produce 150 000 pairs of shoes for the Northern army.

This self-usage helped perfect the machine and therefore supported its rapid dissemination beginning in 1864. Now the machine followed a commodity path; its sales brought learning that led to its further development, which then contributed to sales. Marketing innovations – principally an agency system like Singer's, except with leasing rather than outright sale of the machine – accelerated the sales–learning–invention sequence.

Diversification quickly entered the story. To overcome complementary limitations, McKay bundled his sewing machine with spooling and channelling machines. With success came profits and innovative skills but also market saturation, which provided the means and rationale to diversify. This he did by developing welt-sewing machines, a system of heeling machines, nail-bottoming machines and lasting machines. McKay's diversification relied on and fostered independent invention and new firm formation. For each of these machines, McKay began by developing inventions made by others, merged with other firms, and employed inventors who had earlier developed capital goods for single-product firms. For each, he also spread knowledge that led to the formation of new, innovating firms. And the spread of his machines certainly led back to the form of self-usage that we call learning by doing.

Shoemaking inventions in general shared three features with the process of developing McKay's inventions. First, economic forms often complemented one another. For some inventors, self-usage paved the way for usage as a capital good. Among the 35 surveyed shoe factory managers and foremen – whose occupations prepared them for self-usage – eight had patents used by capital goods firms. Commodity usage and diversification were also interdependent. Of 37 inventors associated with diversifying firms, 13 also gained use with single-product firms, eight of these in firms of their own. Buying basic patents, hiring patentees and merging with single-product firms were all means firms used to diversify.

Second, many new capital goods firms organized their own inventive processes. Based on feedback from users, they acted to overcome technical limitations of earlier machines, to adapt these machines to new uses and to overcome limits in other operations. Diffusion and generation thus were not truly separate activities. Established inventors put such learning to use; the 101 sampled shoemaking patentees who were known to have secured use with capital goods firms averaged 6.5 repeat shoe manufacturing patents (10 times the repeats of other inventors).

Third, the unintended consequences of firm activity were as important for cumulativity as were the intended ones. If learning by inventors already associated with innovative capital goods firms was one side of the learning by selling process, learning by inventors outside these firms was the other. Singer unintentionally fostered invention by Blake, and later McKay unwittingly led to invention by dozens of others, some of whom founded firms that competed with McKay.

New inventors increased when and where sales grew. New shoe manufacturing inventors doubled from the 1860s to the 1880s and remained concentrated in Massachusetts, the centre of shoe and shoe machinery production. The correlation of patenting with the extent of sales, often interpreted as an optimizing response to potential markets, is better understood as an attribute of the learning by selling process. This is shown by the location of sewing machine sales, as indicated by the spatial distribution of company agencies. A large, national and rapidly growing sewing machine market existed by 1860, so that all US residents had an equal interest in sewing machine invention. Yet patenting was located near agencies, changed in location when agencies spread, and concentrated near major sewing machine companies (Thomson, 1987, 1989b, pp. 101–5, 142–4, 168). Clearly, the agencies and their parent companies were loci of learning by selling.

Moreover, contact with established capital goods firms taught others how to organize firms. Many new capital goods firms were formed around new inventions. The number of shoe machine and shoe tool firms listed in the

Boston city directory expanded from eight in the early 1860s to 85 at the century's end. These firms typically introduced new products bearing patents of firm members or independent inventors. That their inventions were from the beginning sold through agency systems was another unintended effect of earlier firms' innovations.

Shoe mechanization would not have occurred, at least at the rate it did, without these unintended consequences. Blake failed to use his machine in his own plant, yet succeeded in gaining use for his machine in half the shoe plants in the country. All the major shoe machines were patented by inventors who were not associated with established capital goods firms. All were at first developed by new firms or for self-usage. Like Blake, virtually all these inventors knew of the practices of capital goods firms, and many formed successful firms by emulating those practices.

Through these forms and their intersection, technical change developed an internal dynamic in which new techniques depended more on earlier innovation than they did on economy-wide changes in income or factor prices. If mechanization were induced by the scarcity of labour or the growth of potential markets, then it hardly would have emerged in the 1850s, when wages stagnated and output growth slowed greatly. Yet mechanization did begin then, and the path it took changed internal technical and institutional conditions so basically that economy-wide changes in demand and skills paled into insignificance by contrast.

ECONOMIC FORMS AND INTERINDUSTRY TECHNOLOGICAL CHANGE

The economic form through which techniques in one industry evolve can also affect technical evolution in other industries. Indeed, together with other factors linking invention among industries – including shared technological principles, occupational structure and locational proximity – economic forms integrate technological change over the whole economy (Rosenberg, 1976; Thomson, 1991).

The prospects for interindustry invention vary among the economic forms. Self-users have the fewest prospects for invention outside the industry, because they have the fewest communication links from which to learn about technical problems and usage possibilities in other industries. Commodity usage eases problems of interindustry technical change. Already tied to capital goods industries, innovators hold widely applicable knowledge of how to invent, make and sell machines and other inputs. Communication and marketing problems for inventions outside an industry, especially when technologies converge, are much less significant than for self-users. Still, communication with users and competitors is far denser than with non-vertically related industries. Diversification faces similar communication problems but provides resources and a rationale for invention outside the industry.

Table 6.2 Crossover patenting by type of shoemaking patent use, 1848–1901 (numbers of patents)

	Self-used	*Commodity usage*	*Diversification*	*Unused*
1 Total patentees	44	64	37	119
2 Crossover patentees	16	36	27	46
3 Shoemaking patents	80	326	431	184
4 Footwear patents	19	11	8	4
5 Crossover patents	32	353	337	557
6 Crossover frequency (2/1)	0.36	0.56	0.73	0.39
7 Crossover extent (5/2)	2.00	9.81	12.48	12.11
8 Crossover rate (5/1)	0.73	5.52	9.11	4.68

Notes: Crossover patentees had one or more crossover patents, and crossover patents were all US patents other than those for shoe manufacturing and footwear.

Sources: US Department of Commerce, *U.S. Patent Classification – Subclass Listing*, class 12 and leather sewing machines from class 112; annual reports of the US Commissioner of Patents; patent grants and city and business directories for 80 US cities.

Four implications for interindustry invention follow. The most immediate is common to all three forms: further invention for the same industry faces fewer barriers than invention for new ones. The reasons are partly technological: convergences and imbalances are more common within industries than between them. There are also social reasons: the connections between an industry and its capital goods firms provide denser flows of technological, marketing and financing information than do the connections between vertically unrelated industries. As a result, technology develops with more autonomy between vertically unrelated industries, and technical change for the whole economy is made up of many relatively separate cumulative processes.

The US shoe industry exemplifies this autonomy, as the patenting behaviour of its inventors shows. The 264 surveyed inventors received 2342 US patents from 1848 to 1901, which Table 6.2 divides into three types: shoe manufacturing, footwear (concerning shoe design, not shoe manufacture), and crossover (those outside the shoe industry). One-half of the 264 surveyed shoe manufacturing patentees took out more than one shoe patent, somewhat above the share who took out *any* of the 32 types of crossover patents (computed from Tables 6.1 and 6.2, lines 1 and 2). Only the category of inventors without any kind of shoemaking use – who thus learned little from shoemaking inventions – had fewer multiple shoemaking patentees than crossover patentees. Moreover, repeat shoemaking invention was far more frequent than any individual type of crossover invention. Whereas 131 inventors took out more than one shoemaking patent, only one-quarter as many invented in the most commonly entered crossover sectors: apparel, metalwares or dry-thread sewing (Thomson, 1991).

Shoe-sewing machine inventors further demonstrate the obstacles to leaving shoemaking. Sixty-three per cent of them took out other types of shoe-manufacturing patents but only 37 per cent patented other kinds of sewing machines. This disparity highlights the importance of economic institutions, which directed invention to shoemaking even though shoe sewing converged technologically less with shoemaking than it did with other sewing machines.

The second implication is that interindustry technical change varies with the economic form taken by technical change within industries: that is, this form affects the frequency, extent and direction of patenting outside the industry.

Self-used invention was most integrated with technical change elsewhere in the economy when techniques of one industry applied widely to others and could be diffused through the operation of labour markets. Craftsmen who made products used in many industries, such as carpenters and toolmakers, often innovated in this way. Likewise, labour markets brought mass production machine tools from firearms to sewing machines.

Shoe producers who used their own inventions did not have widely applicable skills, a broad knowledge of convergences or the ready means to sell in other industries; hence their crossover invention was – when compared to that of inventors gaining use as capital goods – limited and directed to immediately connected patent types. Only 36 per cent of self-users crossed over, by far the lowest share of those with any kind of shoemaking patent use. Those who did cross over averaged but 2.0 crossover patents (lines 6 and 7 of Table 6.2). Their intra-industry inventive orientation is manifested by their 19 footwear patents, which comprised 37 per cent of all of their non-shoemaking patents. In stark contrast, for other forms of usage footwear patents made up only 2 per cent of all non-shoemaking patents.[5]

With their much easier access to other industries, inventors for capital goods firms had more opportunity to invent outside the shoe industry. Most made machinery, and machine innovators could form machinery companies in other industries in much the same way they did in shoemaking. Inventors who gained commodity usage were prolific crossover inventors. The 56 per cent who crossed over averaged 9.8 crossover patents. Inventors for diversifying firms were even more accomplished. They had the most advantages: they were well trained in invention, had access to knowledge of machinery needs and production, and were by occupation overwhelmingly machinists and professional inventors. Almost three-quarters of them crossed over, averaging 12.5 patents outside the shoe industry.

Inventors for diversifying firms also had another advantage: their invention could have been an aspect of the firms' organized efforts to diversify between industries. However, this advantage was little used. Only six shoe machine inventors assigned crossover patents to shoe machinery firms.

These inventions aimed at overcoming imbalances rather than at diversifying product lines. They centred principally on machinery needed to form nails and tack strips used in lasting and nailing machines; as such they constituted a kind of self-usage that was really part of the shoemaking technological system. Three firms diversified among sewing machines and employed prolific generic inventors to enter shoemaking. Only one firm, William Sellers & Company, greatly widened the scope of its product lines; the lasting machine it patented in 1900 added but another machine to its long line of machine tools, steam engines, apparel machines, measuring devices and railway equipment. The much more common course for inventors in diversifying firms was to find different companies to use their crossover patents. Extensive interindustry diversification would wait for the new century.

A third implication is that once a crossover industry was entered, the continuation of invention there depended upon usage within that industry. Table 6.3 examines how capital goods usage of crossover patents related to further crossover invention for 337 inventors (the 264 sampled inventors and another 73 self-users added to get meaningfully large numbers). Crossover inventors with at least one crossover patent used as a capital good (whether with single- or multiple-product firms) averaged 18.7 crossover patents, six times the average for those who did not gain such crossover use. This was more or less true for all forms of usage of shoemaking patents. Hence, usage within the crossover sector, and not the economic form of shoemaking usage, determined continuing crossover invention.

However, the economic form taken by shoemaking inventions did influence whether crossover use was secured. The share who brought crossover patents into use was highest for inventors who gained shoemaking use with diversifying firms (56 per cent in Table 6.3, line 7) and lowest for self-users (17 per cent).[6] Because of common requirements for bringing commodities (and particularly machines) into use, commercial experience in one sector supported success in others. Thus, besides their advantages for crossing over, inventors for shoemaking capital goods firms had advantages in gaining usage for their crossover patents. Consequently, 32 per cent of them gained crossover usage, whereas only 6 per cent of those with self-used shoemaking patents did.

The final implication integrates the economic form with the content of invention. All capital goods inventions did not equally support crossing over: machinery tied together technical change among industries far more than did lasts and shoe tools. For machine shops were loci of interindustry communication. They often made machinery for more than one industry and employed workers who, in prior jobs, had made machines for other industries. Inventors of many kinds of products came to machine shops to have their creations perfected and produced. By contrast, lasts and shoe tools only reached a few sectors.

Therefore, shoe last experience could apply to other uses of the pattern lathe, such as gun stocks, axe handles and hat blocks, but technical knowledge of shoe machines applied to machines and mechanisms everywhere in the

Table 6.3 Crossover patenting, by shoemaking and crossover use, 1848–1901 (numbers of patents)

	Self-used	*Commodity usage*	*Diversification*	*Unused*
With crossovers used as capital goods				
1 Number	7	17	15	24
2 Crossover patents	90	297	311	480
3 Extent (2/1)	12.86	17.47	20.73	20.00
With unused crossovers				
4 Number	34	19	12	22
5 Crossover patents	115	56	26	77
6 Extent (5/4)	3.38	2.95	2.17	3.50
7 Share used [1/(1+4)]	0.17	0.47	0.56	0.52

Notes: Inventors are considered to have had crossover patents used as capital goods when they owned, managed or assigned at least one patent to capital good firms making goods that could use the crossover patent. The data base differs from that of Tables 6.1 and 6.2 because 73 additional self-using inventors (25 with crossover patents) were added in order to make numbers with crossover patents used as capital goods large enough to be meaningful. This larger sample has the same crossover frequency as self-users in Table 6.2, but has a considerably higher extent. All the conclusions derived Table 6.2 would also be supported by the larger sample here.

Sources: US Department of Commerce, *U.S. Patent Classification – Subclass Listing*, class 12 and leather sewing machines from class 112; annual reports of the US Commissioner of Patents; patent grants and city and business directories for 80 US cities.

economy. Machines formed 71 per cent of shoemaking patents and 62 per cent of crossover patents, and the disproportionate concentration of shoe machine inventors on crossover machinery patents demonstrates that these two spheres of machinery invention were mutually supporting (Thomson, 1991). Hence inventors who gained use with shoe machinery firms crossed over twice as frequently as those who gained use with last or shoe tool firms (69 to 33 per cent).

Machinery *as an occupation* offered particularly strong support for crossover invention. Shoemaking inventors came from variety of occupations; one-half made shoes and shoe inputs when they pursued their first shoe patents, and the remainder were generic machinists, professional inventors and workers of iron, leather, wood and many other materials. Machinists and the professional inventors to whom they were closely allied (including draughtsmen, modelmakers, patent solicitors and full-time inventors) used skill and communication advantages to invent more. They averaged twice as many shoemaking patents as other inventors, and over three times as many crossover patents. Mainly, this gap represents their greater ability to gain capital-goods usage; 58 per cent gained such usage for shoemaking patents and 62 per cent for crossover inventions, both doubling the shares of other occupations. With greater opportunity to cross over, machinists and professional inventors did so in higher proportions than did other occupations (64 to 37 per cent). The machinery industry arguably led both shoe-making and crossover invention through the invention of machinists and the supportive institutional conditions that it provided for all inventors (Thomson, 1990).

Occupational differences in inventiveness overlaid a new dimension on to the basic structure of technical change's economic forms. Although machinists had advantages, inventors from all occupations crossed over more or less successfully because of knowledge communicated through the different forms of usage. The difference was particularly pronounced for non-machinists. Ten of the 42 who gained shoemaking use as capital goods also secured crossover usage, while only eight of the 121 without shoemaking capital goods usage secured crossover usage. Machinery integrated industries, especially when it diffused through sale.

ECONOMIC STAGES OF TECHNOLOGICAL CHANGE

All good history is attuned to the question of qualitative change. Historians frequently interpret technological change as a source of economic stages, but they rarely turn the problem on its head and examine the economic stages of technical change. The economic forms of technological change help to do just that.

The possibility of periodizing is suggested by the sequence of McKay's machines, moving from Blake's self-usage to McKay's commodity usage and then to McKay's diversification. The usefulness of periodization for understanding technical change comes when stages have their own dynamics and when these dynamics foster transitions to later stages. The cumulative character of technological change makes it possible to think of stages in this sense for, in any of the economic forms, technological change does not simply accommodate to the state of the economy but develops instead through its own use–learning–invention feedbacks. Different feedback processes can distinguish stages, and earlier forms can create conditions for and paths towards later forms.

This section will investigate whether the economic forms structured stages of technological change in shoemaking. Because technological change involves cumulative processes within particular sectors and connections of those processes among sectors, this investigation must look both within shoemaking and at its integration with the broader economy.

Within Shoemaking

Stages cannot be a simple progression from one form to another; all three forms coexisted in the late nineteenth century and continue to do so today (von Hippel, 1988). However, stages can be distinguished by the dominant economic form. Self-usage characterized the innovators of the craft period, which lasted into the 1850s. Typically, shoemakers improved craft techniques for use in their own shops. Techniques spread mainly through the movement of workers trained in these shops. Means of production, such as tools and pegs, evolved in part through self-usage. Craft self-usage led to information flows in eastern Massachusetts that helped develop the putting-out system, the pegged shoe and craft technique. It also supported new product development by input producers making pegs, lasts and shoe tools. However, because handicraft skills were transmitted through training in the production process and limited the productivity impact of other inputs, commodity usage could only be of secondary importance for changing techniques in this period (Thomson, 1989b).

Commodity usage was the predominant form bringing shoe mechanization. Blake designed his sewing machine for use in his own factory, but he soon discovered that self-usage could not overcome the barriers of inadequate technological and production skills, limited access to finance, and a scale too small to meet development expenses. Organized to overcome such limits, new machinery firms in the third quarter of the century not only sold or leased machines, but also spread knowledge that supported their continuing invention, invention by others and the proliferation of shoe machinery firms.

Self-usage continued as learning by doing, now dependent on capital goods firms and their products. Shop training still supplied shoemaking skills, but shoe machinery firms also trained workers and spread shop-floor innovations to other shops. Shoemakers continued to develop new equipment; even as machinery spread and grew more sophisticated, shoemakers formed a steady 40 per cent of those entering shoemaking patenting from 1860 to 1901. Similarly, self-using inventors formed about the same share of new inventors after 1880 as they had before. Commodity usage, then, could not actually replace self-usage, for the more successful commodity usage became at widely selling new capital goods, the more it promoted secondary invention by self-users. But successful secondary inventions were often channelled back into the ever-wider current of commodity usage; ongoing self-usage thus supported the domination of the commodity form.

Diversification followed on the success of commodity usage, and in some cases was a condition for that success. It often began a decade or two after the capital goods firms introduced their first products. Hence dry-thread sewing machine producers and Townsend began to diversify in the 1860s, McKay in the 1870s, and Goodyear in the 1880s. Diversifying firms often dominated and expanded their product lines until, when United Shoe Machinery was formed in 1899, one firm nearly monopolized bottoming and heeling machines. By the 1880s and 1890s, technical change was no longer predominantly associated with vertical disintegration and new firm formation. Rather, ongoing technical change came with growing integration vertically into marketing and product refinement, horizontally via competition and mergers, and laterally into new products and new markets (Thomson, 1989a).

Still, diversifying firms did not eliminate other forms of usage in the third stage of shoemaking technological change. Typically they bought out patentees or firms in lines they wanted to enter, and their own sales occasioned learning and entry by other firms. Their very success brought about learning at the point of production. And, as McKay's tack strip inventions show, diversifying firms also invented for their own use. Far from dying, self-usage and commodity usage provided fundamental support to the increasingly powerful diversifying firms.[7]

The periodization gains usefulness because the dynamics of earlier stages helps to account for the birth of later stages. Self-usage created conditions for commodity usage. Self-used innovations in the craft period increased the profitability and growth of Massachusetts shoe firms such that many achieved a scale needed to introduce shoe machinery. Had this scale not been achieved, learning by selling would not have added its decisive cumulativity to technical change. Vertical disintegration helped form separate crafts of lastmaking and shoe toolmaking, and innovators from all these crafts advanced shoe mechanization.

Commodity usage led to diversification through the logic of the product cycle. After a point, retained earnings could not be reinvested profitably in the same commodities. Firms then employed advantages they had gained in earlier innovation to develop new commodities that would sustain their accumulation.

However, the processes of and transitions between stages cannot be understood within shoemaking alone, as the transition to the stage of commodity usage illustrates. Until mechanization overcame the limitations of craft skill, commodity usage could not fully develop. But earlier shoe innovation cannot by itself account for mechanization. Shoe and shoe input crafts did not invent many shoe machines, and neither did they produce these machines, generate the pattern lathes that standardized lasts or develop machine agency systems that raised sales beyond custom markets. Stage shifts of technological change cannot be understood by focusing solely on the path of craftsmen making shoes and their traditional inputs. Industries, like people, are not islands.

Shoemaking and the Broader Economy

Both mechanization and commodity usage in the shoe industry depended on technological change elsewhere in the economy, and hence on the economic forms, stages and history of such change. Because this was true of many industries, stages of technological change took on an economy-wide character.

Many kinds of craft production evolved through self-usage in the first half of the nineteenth century. Mechanization also developed through self-usage. Many of the machines of the British Industrial Revolution evolved in this way. Until the second quarter of the nineteenth century, the American cotton textile industry developed through the same process. Even mass production machine tools spread through self-usage until the Civil War.

Unlike most traditional crafts, self-used machinery integrated many sectors of the economy. Most crafts evolved with substantial autonomy from one another, so that the economy was effectively partitioned into separate paths of technical change. But, as early nineteenth-century labour markets for machinists indicate, machinery inherently applied to more than one trade. Especially in industries with large plants and high capital requirements, the self-usage form could generate machinery and, through mobility and sometimes licensing, spread technological principles within and between industries (Roe, 1916; Rosenberg, 1976; Smith, 1977; Hounshell, 1984; Cooper, 1991).

As machinery users multiplied in the second quarter of the century, a machinery industry was born, and the commodity form of mechanization grew in importance. Initially, many machines were custom ordered, and their sale resembled more the luxury shoe trade than wholesale shoe marketing. Even so, sales spread machinery (or, through licensing, the right to use technological principles) within industries, and machinists used their technological, production and marketing skills to develop new machines, establish new firms and enter new markets.

Only within this context can shoe mechanization be understood. The machinery industry supplied not only skills and personnel, but also the economic form through which shoemaking was mechanized. Starting with the sewing machine, machinists and shoemakers in contact with machinists invented and formed shoe machinery firms to market their discoveries. Although their small size and limited mechanical knowledge discouraged invention for self-usage, shoe manufacturers were quite willing to purchase machines developed by others. New shoe machine firms thus brought to shoemaking the machinery and the stage of commodity usage that the shoe industry could not supply for itself.

Many industries mechanized in similar ways; to a significant extent, it was the machinery industry's broad reach that made technical change an economy-wide process in the nineteenth century. Machinery generated inventors who moved into many industries. Some inventors failed, but others successfully forged new paths down which firms developed not only techniques, but also markets, marketing systems and managerial structures. These paths gained autonomy from the generic machinery sector that nurtured them, but they continued to call on that sector for inventors and producers. In return, these paths contributed technical principles, inventive personnel and novel marketing and managerial institutions that advanced mechanization elsewhere. Stages of technological change thus pertained to the whole machinery-selling and using sector, and the transition to commodity usage was also a transition to a more technologically integrated economy.

Shoe mechanization took from and gave to other sectors. It drew on other sectors for inventive and production talent, the sewing machine and other applicable inventions, mass production machine tools, and novel business practices such as Singer's marketing system. It gave back inventions applicable to a few sectors and inventors who crossed over into many sectors. The agency and leasing systems that allowed Singer, McKay and Goodyear to produce and distribute standard models were copied by others, breaking with traditional machine-shop practice and scale (Chandler, 1977, 1990). The sales of such machinery firms spread knowledge much more quickly, widely and thoroughly than had the movement of workers or licensing, and the growing scale of these firms helped transform mass-production metalworking machinery into commodities, setting them on the path of commodity usage.

Not until the twentieth century did diversification tie together technical change among industries. In the nineteenth century, firms developed new products within their main line of machinery, and only developed other equipment when needed for their own production processes. The diversification of the new century would supplement learning among firms in different industries with interindustry, intrafirm planning. In the new stage, technological change was then organized by firms to apply a core technology to all its potential markets (see Chapter 3).

Economic forms of technological change played critical if neglected roles in capitalist development. They affected the pace of technological change within industries, the interrelation of technological change among industries, and the birth of new economic forms. They also had broader effects. Technical change plays a central role in most interpretations of economic development by affecting competitive structure and industrial concentration, income growth and distribution, profit and accumulation rates, class structure, penetration into non-capitalist systems and, through these factors, the capacity and incentive to undertake technical change. All these effects varied with the economic organization of the process of technological change: that is, they followed not simply from technological advance but also from the economic form taken by that advance. The results of technological change were thus inseparable from the socio-economic process that brought this change about. If technological change was an engine of growth, economic forms were engine types, and their differences led to growth of different kinds, rates and consequences for further technical change.

Notes

1. My concern is with inventions used by firms. Some new products were also consumption goods, notably the sewing machine, and much of the analysis would also pertain to these goods. For the distinction among types of usage, see Thomson (1989b). Needless to say, these three types do not exhaust even firm-connected technical changes. Techniques can be diffused by licensing or patent

assignment, by sharing among firms, by such publications as patents, journals and technical dictionaries, by observation in industrial expositions or in family and community gatherings, and by the state and its contracting policies.

2. Many barriers to self-usage can be overcome if firms share technology. Then the smallness of firms is not so debilitating, and diffusion is obviously eased. Still, technology sharing is not easy to implement or to police; it works best for incremental changes within closely-knit groups of firms (Allen, 1983).
3. Assumptions concerning self-usage are least secure. Self-used inventions are, for statistical purposes, nearly invisible. They leave no evidence in the products of capital goods firms or, if strictly self-used, in assignments (other than to the self-using firms). Furthermore, self-users may have patented less because they preferred secrecy to patent protection. (On these issues, see Thomson, 1989b, pp. 189–94.) Assignments are limited to those at the time of the patent grant. On the methods and benefits of analysing later assignments, see Cooper (1987 and 1991).
4. The only anomaly here is the similarity of self-users and non-used. This is largely the result of the higher share of non-shoemakers among those not gaining use; if attention is restricted to shoemakers, inventors with self-usage averaged over twice as many repeat patents as did inventors not gaining use. On ties among occupation, usage and repeat inventing, see Thomson (1989b), pp. 186–94. Rates of repeat invention here are almost three times those in Table 12.4 of that work, a difference explained because this study included shoe-sewing machines and shoe manufacturing patents not categorized as such by the Patent Office, and restricted its attention to inventors listed in city directories (who gained use more frequently than did others).
5. The limited rationale for self-users to patent is less applicable to crossover patents, which could not readily be self-used without changing lines of business.
6. Considering forms of technical change outside shoemaking makes sense of the otherwise anomalous experience of those without shoemaking use, whose rate of crossover patenting (4.7 in Table 6.2) was far higher than the 0.7 of self-users and approached the 5.5 of commodity users. Many of these inventors gained use and invented widely in other industries; they entered shoemaking in an exploratory way, but found their calling elsewhere. For them, use outside shoemaking determined their patent histories.
7. Of course, stages of technical change were not independent of wider economic forces, for technical change in all stages rested on usage, and hence on determinants of diffusion such as the size of firms and markets and the availability and cost of finance and labour. Still, past technical change, by fostering firm and market growth and lowering unit labour costs, helped shape these factors and the limits they posed to present technical change.

References

Allen, R.C. (1983) 'Collective Invention', *Journal of Economic Behavior and Organization*, 4 (March).

Chandler, A.D. Jr, (1977) *The Visible Hand: The Managerial Revolution in American Business* (Cambridge, Mass.: Belknap Press).

—— (1990) *Scale and Scope: The Dynamics of Industrial Capitalism* (Cambridge, Mass.: Belknap Press).

Cooper, C.C. (1987) 'Thomas Blanchard's Woodworking Machines: Tracking 19th-

Century Technological Diffusion', *IA: The Journal of the Society for Industrial Archaeology*, 13.

—— (1991) *Shaping Invention: Thomas Blanchard's Machinery and Patent Management in Nineteenth-Century America* (New York: Columbia University Press).

Hounshell, D. (1984) *From the American System to Mass Production, 1800–1932* (Baltimore Md.: Johns Hopkins University Press).

Levine, D.P. (1981) *Economic Theory*, vol. 2 (London: Routledge & Kegan Paul).

Roe, J. W. (1916) *English and American Tool Builders* (New Haven, Conn.: Yale University Press).

Rosenberg, N. (1976) 'Technological Change in the Machine Tool Industry', *Perspectives on Technology* (Cambridge: Cambridge University Press).

Shapiro, N. (1986) 'Innovation, New Industries, and New Firms', *Eastern Economic Journal*, 12 (January–March).

Smith, M.R. (1977) *Harpers Ferry Armory and the New Technology* (Ithaca, NY: Cornell University Press).

Thomson, R. (1986) 'Between Invention and Practicality: The Development of the Sewing Machine', unpublished manuscript.

—— (1987) 'Learning by Selling and Invention: The Case of the Sewing Machine', *Journal of Economic History*, 47 (June).

—— (1989a) 'Invention, Markets, and the Scope of the Firm: The Nineteenth Century U.S. Shoe Machinery Industry', *Business and Economic History*, 18.

—— (1989b) *The Path to Mechanized Shoe Production in the United States* (Chapel Hill, NC: University of North Carolina Press).

—— (1990) 'Machinery as the Invention Industry', unpublished manuscript.

—— (1991) 'Crossover Inventors and Technological Linkages: American Shoemaking and the Broader Economy, 1848–1901', *Technology and Culture*, October.

von Hippel, E. (1988) *The Sources of Innovation* (New York: Oxford University Press).

7 A Model of the Productivity Gap: Convergence or Divergence?[1]

Donald J. Harris

INTRODUCTION

A notable feature of the post-war experience of productivity growth is a tendency to convergence in productivity levels among a selected sample of advanced capitalist economies. This tendency has been identified and discussed by a number of observers.[2] However, the strength and generality of this tendency are a matter of dispute.[3] Even if it is accepted for a specific subsample of countries, it remains evident that there is great diversity in the actual pattern of experience of a wider class of countries, including the less developed countries, observed over the same period. Among this wider class, one finds the coexistence of both convergence and divergence, with no clear and unambiguous case for either tendency to prevail across the whole set of countries.[4] This result indicates that the picture is more complex than appears at first sight and, correspondingly, calls for a deeper investigation.

Viewed over a longer time period, the overall picture becomes even more varied and complex. In this connection, it is instructive to examine the data recently assembled by Angus Maddison (1987) for six advanced capitalist economies. The relevant results for this sample, reproduced here in Table 7.1, attest to the remarkable performance of the five follower countries in reducing the productivity gap relative to the USA by almost three-quarters (from 59 to 16 per cent) in the period 1950–84. But what is equally striking is that by 1973 the five countries had just barely succeeded in restoring the relative position which they occupied a century before and had lost in the intervening period.

Thus, what we find over this longer period for this particular sample of countries is an *alternation of two phases of productivity growth*, constituting a kind of long-term cycle. The recent phase of this cycle, the one that has captured the attention of observers, clearly exhibits a tendency to convergence. But, evidently, this phase is preceded by one in which there occurs a widening gap or divergence in productivity levels by a substantial margin of about 30 percentage points extending over a 80-year period. It is also noteworthy that, by the end of this long period of 114 years, the USA (which has been the leader throughout most of the period, even though its

Table 7.1 Comparative levels of productivity (GDP per hour worked), 1870–1984 (US GDP per hour worked = 100)

	1870	*1913*	*1950*	*1973*	*1984*
France	54.7	49.2	41.5	74.7	97.5
Germany	58.9	55.7	33.6	72.8	90.5
Japan	19.3	18.0	13.9	43.8	55.6
Netherlands	105.2	74.6	56.5	87.5	97.2
UK	110.9	80.0	58.6	68.8	80.6
Five country (weighted) average	70.5	56.2	40.9	69.5	84.3
USA	100.0	100.0	100.0	100.0	100.0

Source: Maddison (1987), p. 651. Reproduced with the permission of the editors of the *Journal of Economic Literature*.

leadership edge has been reduced) still retains a commanding lead of about 16 percentage points.

In considering the historical record of productivity growth, one needs to take into account also the changing relative position of the UK which, after all, was the leader in the nineteenth century. Available evidence indicates that the UK loses its leadership position to the USA around 1880, and is gradually overtaken in the twentieth century by a succession of countries: Sweden, France, Germany and Italy (Matthews, Feinstein and Odling-Smee, 1982, p. 32). Despite its impressive recovery in the period 1950–84, it still lags behind the USA by almost 20 per cent in 1984. Over the long haul, then, there occurs an overtaking effect and a change in leadership with respect to the position of the UK. This aspect of the matter adds further complexity to the general picture. One may well wonder, if recent experience is projected forwards, whether another such episode of changing leadership is now in the making as regards the leadership position of the USA.

A final point concerns the relative position of the less developed countries. For this comparison, owing to the absence of comprehensive productivity data, one must rely on per capitum income figures, which admittedly provide only a rough guide to productivity levels. The data (Harris, 1986; World Bank, 1986, pp. 180–1) show that, for much of the post-war period, the group of 'middle-income economies' have narrowed the gap relative to the top group of 'industrial market economies'. For the 'low-income economies' as a group, on the other hand, the gap has actually been increasing relative to the top. As to the actual magnitude of the gaps involved, the ratio between the top and the bottom stands at 44:1 in 1984; between the top and the middle it is 9:1.

This record of experience in productivity growth poses deep problems for economic analysis.[5] The overall picture is evidently much more diverse and complex than either a simple convergence thesis or its opposite (a divergence and polarization thesis) would suggest. This makes the analytical problems even more difficult and less straightforward. However, it is possible and necessary to approach these problems in a step-by-step manner. Accordingly, in this chapter a specific set of analytical questions that this picture raises will be addressed. The focus will be particularly on the nature of the so-called convergence process as such, asking the following questions: under what circumstances does the process of productivity growth tend to converge or to diverge? What are the factors that determine such convergence or divergence? If a productivity gap persists, what determines its ultimate size?

A HEURISTIC MODEL

For the purpose of providing an answer to this set of questions, below is constructed what might be called 'a model of a productivity race' in which there are specified relationships governing the rate of productivity growth among different production units viewed as countries or regions. From these relationships one can find certain characteristic conditions, related to the parameters of the productivity-increasing process, which allow a direct inference concerning the factors that determine the possibility of convergence/divergence among different units and the size of the gap, if any, that remains between them.

It should be remarked that this is a model of pure productivity growth. It abstracts essential features of productivity growth as an endogenous process, putting aside other factors that are usually considered to affect growth of output such as saving/investment rates, aggregate demand, supply of labour and natural resources. The advantage of this particular model is its ability to capture, in a simplified manner, some of the essential properties of what is presently known, from the most advanced work on technological change, about the nature of technological change as an endogenous process. These ongoing efforts in the direction of understanding and conceptualizing technological change offer rich possibilities for getting a firm grasp on the kinds of factor that account for differences in productivity experience across different countries.[6]

So far as the concept of productivity used here is concerned, it is the simple and well-defined concept of labour productivity: that is, average product per unit of labour. This bears comparison with the neoclassical concept of total factor productivity, measured as the ratio of output to a weighted index of (augmented) capital and labour inputs. All the well known capital-theoretic problems implicit in the aggregate production function underlying that concept are avoided here by focusing on labour productivity. Actually, in this model it is assumed that labour is the only input in production, although there is an augmentation effect on the side of labour arising from experience. Correspondingly, factors related to 'capital deepening' that have traditionally been used to account for productivity growth, whether one thinks of capital deepening either as increasing mechanization or as variation of the length of life of different vintages of capital, are left out of consideration. Other factors, such as investment in human capital, are also ignored.

It is assumed that productivity increase is *a self-generating process*. This self-generating feature derives from two considerations that are crucial to the model. First, it derives from the operation of what might be termed the knowledge industry,[7] which consists of the congeries of activities taking place within the universities and research institutes, within the R&D divisions of firms, in industrial laboratories, and in the activities of people tinkering in the basement. It therefore includes what is commonly referred to as R&D activity, but much else besides.[8] Operationally, the output of this industry is embodied in technical blueprints, patents, professional and trade journals, books, videos, computer software and so on. These outputs are linked to production, and hence to productivity, in many complex ways that defy detailed specification. Nevertheless, that link is clear and well established.[9] Conceptually, what this analysis seeks to capture is the crucial role of these activities as a determinant of overall productivity growth.

Second, the self-generating feature of productivity growth derives from an intrinsic characteristic of the production process, namely that experience counts in some meaningful sense. In particular, it counts here towards further increase in productivity. In this respect, there exists a 'learning effect', which is modelled here as both a learning-by-doing effect and a learning-by-using effect. This feature of the model also conforms well to ideas that have been demonstrated and documented in the literature.[10]

This particular way of approaching the problem of productivity growth has the significant implication that every producing unit (country or region) has the capacity to generate its own productivity growth from its own learning, subject to a critical threshold effect that, as we shall see, may operate to inhibit some units from starting up the process. Every producing unit has, so to speak, the capacity to pull itself up by its own bootstraps, provided that the required minimum condition is met. Therefore, observed differentials in performance among units, instead of being reduced simply to arbitrary external factors, barriers or limits, must be accounted for by factors that are internal to the productivity increasing process. Furthermore, once we know what these factors are, we can say under what conditions the process will tend to converge or to diverge and what determines the asymptotic state of that process as regards the magnitude of the productivity gap. This is essentially the thrust of the analysis presented here.

For the purpose of this analysis, the following assumptions about knowledge as a commodity will be made.[11] First, it is permanent and indestructible, and hence does not depreciate over time. Second, it is a produced commodity, produced by its own production process in the knowledge industry. Third, it has the capacity to increase the productivity of all industries including its own; hence it is a high-powered commodity in this sense. Fourth, it generates significant externalities in the course of its production and use, in the strict sense that any producer can benefit from access to and use of a given total quantity of knowledge without diminishing the amount available to others. This externality feature of knowledge as a commodity implies that there are intrinsic problems of establishing property rights and hence of appropriation of income from its use. Thus the idea of inferring a unique market-determined price of knowledge or an immediate connection with income of its owners is highly problematical. For this reason, it is worth emphasizing that no significance is assigned here to the pricing and income distribution side of the production of knowledge.

THE MODEL

Proceeding now to construct the relations of the model, let us define a critical variable x, the stock of knowledge, which is the sum of all the flows of knowledge generated in the past. Thus

$$x = \int_0^{\tau} \dot{x}\, \mathrm{d}t \tag{7.1}$$

It would be straightforward to extend this formulation to allow for depreciation of the stock of knowledge, but this complication is not considered here. The essential point is that x is assumed to be a scalar. This notion of an aggregate of knowledge is used here for heuristic purposes only. A simple way of giving it a concrete representation is, for instance, as a number of blueprints or a number of patents. There are, of course, important theoretical and practical problems involved in constructing such an aggregate (as with many other aggregates commonly used in economic analysis) in a real-world context of heterogeneous knowledge commodities, but these problems are not considered here, and neither are they strictly relevant for present purposes.

The production characteristics of the knowledge industry are specified as follows:

$$\dot{x} = f(x, L) = \phi(x).L \tag{7.2}$$

Here, the flow output of knowledge, $\dot{x}$, is a function of the stock of knowledge x and the labour input L. The stock of knowledge represents an index of productive experience, which has a positive effect on production through a process of learning-by-doing. The learning function is further specified to be a function $\phi(x)$ which is a multiplicative factor applied to the labour input. This formulation says simply that the track record of experience in producing knowledge, as measured by the cumulated stock of knowledge already produced, governs the productivity of labour in producing knowledge.

$$\frac{\phi(x)}{x} > \phi' > 0 \tag{7.3}$$

It is assumed that the average product of experience is greater than the marginal product and that the marginal product is positive. Thus there is a kind of 'diminishing returns' to experience. This assumption is intended to capture an idea that recurs in the literature, taking different forms. In its most common form it is the idea of running up against a frontier of technological knowledge, which essentially implies that beyond a certain point the yield of incremental efforts in R&D activity rapidly falls off (to zero in the extreme). It is sometimes tied to 'Wolff's Law', referring to a general tendency to 'retardation of progress' (Freeman, 1982, p. 216). It could also be derived from the idea of a 'lock-in effect' arising from cumulative experience along a given trajectory of technological development (Dosi, 1984). Or it could be that there is a kind of 'dead weight' of past experience connected with the social and institutional structures that it generates, such as 'the accumulation of special-interest groups' (Olson, 1982). Whatever form it takes, this idea evidently entails the existence of some condition within the knowledge-producing industry that acts cumulatively to retard the process of increase in productivity.[12] That condition may itself be considered to be of an essentially transitory nature if, over time, 'major breakthroughs' in knowledge occur so as to expand the scope for productivity increases at any level of experience. Nevertheless, while recognizing its 'short run' character in this sense, the analytical implications of this idea are worth exploring.[13]

$$L \geqslant L^* \tag{7.4}$$

It is assumed, further, that in order to start up the knowledge industry, it is strictly necessary to have some positive amount of labour input to begin with. Thus there is a kind of critical mass, or minimum threshold, of engineers, physicists, economists and so on that has to be assembled in order to run an effective knowledge-producing process. This assumption also captures an idea that is commonly found in the literature on R&D. It has the significant implication that any unit (country or region) which is unable, for whatever reason, to mount the required minimum scale of the activity is unable to gain the full advantages of the productivity-increasing process.

Now, assume that there is a second productive sector, the y-sector, that produces a consumption commodity. Output of this commodity, y, is produced by labour, L_y. The labour employed in this sector is able to enhance its productivity by drawing on the total stock of knowledge accumulated from production of the x-sector without diminishing the amount of it available to that sector. The same total stock of knowledge therefore enters into the production equation of both the x- and y-sector. In the x-sector, however, it represents a *learning-by-doing effect*, whereas here, in the y-sector, it incorporates a *learning-by-using effect*. This learning-by-using effect is specified to be a multiplicative factor applied to the labour input. Thus we have:

$$y = h(x, L_y) = \mu(x).L_y \tag{7.5}$$

We thus have here a two-sector economy, with a knowledge-producing and knowledge-using (consumption-good producing) sector. There is a degree of circularity in production insofar as the knowledge output re-enters the productive process as the stock of experience, giving rise to learning effects in both sectors. There is an externality feature of knowledge associated with the fact that both sectors draw on the same total stock of knowledge to boost their productivity. Output of the consumption commodity, although forming part of the aggregate national income, drops out of the picture when viewed from the standpoint of the total reproductive process. In the subsequent analysis, no attention is given either to consumption behaviour or to movements of the aggregate national income: the focus is entirely on the production side, specifically on productivity growth which is uniquely connected with growth in the stock of knowledge.

Now, assume that there are two countries (or regions), A and B. Both have an established and viable knowledge-producing industry and a consumption-good industry.[14] Production conditions are the same in both countries.[15] Country A is the leader in the strict sense that it has a greater stock of knowledge than country B, so that $x_A > x_B$. Correspondingly, A also has allround higher levels of labour productivity. In addition, country A allocates relatively more labour to the knowledge industry than country B, so that $L_A > L_B$. Insofar as there exists a gap in the stock of knowledge between country A and B, there is room for a one-way process of diffusion of knowledge from A to B. Assume that diffusion itself is costless in terms of labour and that the amount of knowledge transmitted to B at any moment is proportional to the size of the gap by a factor of proportionality equal to δ. Accordingly, we have the following equations of production of knowledge in both countries:

$$\dot{x}_A = \phi(x_A)L_A \qquad x_A < x_B$$

$$L_A \geqslant L_B \tag{7.6}$$

$$\dot{x}_B = \phi(x_B)L_B + \delta\,(x_A - x_B) \quad 0 \leqslant \delta < 1 \tag{7.7}$$

A convenient interpretation of the diffusion term in (7.7) is that it represents a direct transfer from A to B that is costless to both A and B. It amounts, therefore, to a kind of 'spillover effect' or pure externality. The parameter, δ, could then be taken as a measure of absorptive capacity in B, hence dependent on internal conditions within country B (such as range and depth of social infrastructure, size of the market, language skills and policies of the national state); or δ could be a reflection of regulative measures and other institutional barriers in A to the export of knowledge. An alternative interpretation is that the diffusion term represents a flow of foreign investment from A to B; but this interpretation would raise further complications that cannot be pursued here. Whatever the case, it is supposed that this transfer has a direct impact on the current flow output of knowledge in B equivalent to the size of the transfer. The impact is assumed to be positive;

but one could introduce the possibility that it is negative because of the existence of retarding effects from the transfer process.[16]

In practice, of course, there are likely to be significant resource costs of adoption of imported knowledge and of adaptation to local conditions. Insofar as these are accountable to labour costs, they can conveniently be absorbed into L^* for the importing country. A more complex treatment, consistent with the spirit of this model, would be to make diffusion itself a labour-using activity subject to its own learning process. This is a possible extension of the model.

The analytical problem that is posed now is the following. If both countries operate in accordance with the conditions specified in this model, what would be the associated pattern of productivity growth over time, and what is the long-run outcome of the process as regards the size of the gap in productivity levels? Since productivity levels in both countries are uniquely related to the prevailing stock of knowledge, the analysis focuses on movements in this variable.

DYNAMICS OF THE PRODUCTIVITY GAP

Equations (7.6) and (7.7) constitute the key dynamic relationships, indicating how the two countries evolve over time, starting from given initial conditions. To simplify the analysis and sharpen the results, let the learning function in both countries conform to the following linear relationship:

$$\phi(x_i) = a + bx_i, \quad i = A, B; a > 0, b > 0 \tag{7.8}$$

Then, by transforming (7.6) and (7.7) to proportional rates of growth and subtracting, we get

$$g_A - g_B = \frac{aL_B}{x_A}\left(\frac{L_A}{L_B} - \frac{x_A}{x_B}\right) + b(L_A - L_B) - \delta\left(\frac{x_A}{x_B} - 1\right),$$

$$g_i = \dot{x}_i/x_i, \; i = A, B \tag{7.9}$$

For clarifying the properties of the underlying process, we can distinguish the following cases.

Case 1: $\delta = 0, L_A = L_B, x_A > x_B$

Here, A is the leader in the stock of knowledge, but the two countries are equal in every other respect, and there is no diffusion. In this case, equation (7.9) simplifies to

$$g_A - g_B = \frac{aL_A}{x_A}\left(1 - \frac{x_A}{x_B}\right) < 0 \tag{7.10}$$

Since $g_A < g_B$, the ratio x_A/x_B falls. There is a process of convergence to a steady state. However, the stocks of knowledge are never equalized; they diverge in absolute terms. The speed of convergence is determined by aL_A/x_A which reflects the role of diminishing returns to experience in A. In particular, a/x_A is the difference between the average and the marginal product of experience and it diminishes as experience grows. This result indicates that what dominates the process of convergence is diminishing returns to experience in the leading region. Thus it appears that the leader leads not only in experience; it also leads the process of convergence by its slowing down from 'ageing' or 'maturing' of experience.

Case 2: $1 > \delta > 0, L_A = L_B, x_A > x_B$

This case allows for diffusion from A to B. Equation (7.9) now becomes

$$g_A - g_B = \frac{aL_A}{x_A}\left(1 - \frac{x_A}{x_B}\right) - \delta\left(\frac{x_A}{x_B} - 1\right) < 0 \tag{7.11}$$

Here again, $g_A < g_B$, the ratio x_A/x_B falls, and there is convergence in growth rates but not in absolute terms. The speed of convergence is augmented in this case by the existence of diffusion from A to B. Contrariwise, if $\delta < 0$, implying negative spillovers, it is easy to see that there is no convergence; x_A/x_B rises without limit.

Case 3: $1 > \delta > 0, L_A > L_B, x_A > x_B$

This is the general case, encompassing full differentiation among countries and diffusion of knowledge betweem them. The basic story which can be told in this case is as follows. For L_A/L_B sufficiently large in relation to x_A/x_B, country A has an advantage deriving from its larger allocation of labour to the knowledge industry. This advantage allows it to grow faster than B, so that x_A/x_B increases and, correspondingly, the productivity gap increases. However, part of this advantage, as represented by the first term on the right-hand side in equation (7.9), is diminished by growing experience (due to diminishing returns to experience) as x_A rises both absolutely and relatively to x_B. It is converted to a disadvantage as x_A/x_B comes to exceed L_A/L_B. This advantage is diminished also by the increasing contribution (represented by the third term on the right-hand side of equation (7.9)) that the growing gap in the stock of knowledge makes to growth in B due to diffusion of knowledge from A to B. Both these factors contribute to reducing the difference in growth rates between A and B. Consequently, the magnitude of the gap in the stock of knowledge, while continuing to grow, approaches an upper boundary given by the critical ratio:

$$(x_A/x_B)^* = \frac{b}{\delta}(L_A - L_B) + 1 \tag{7.12}$$

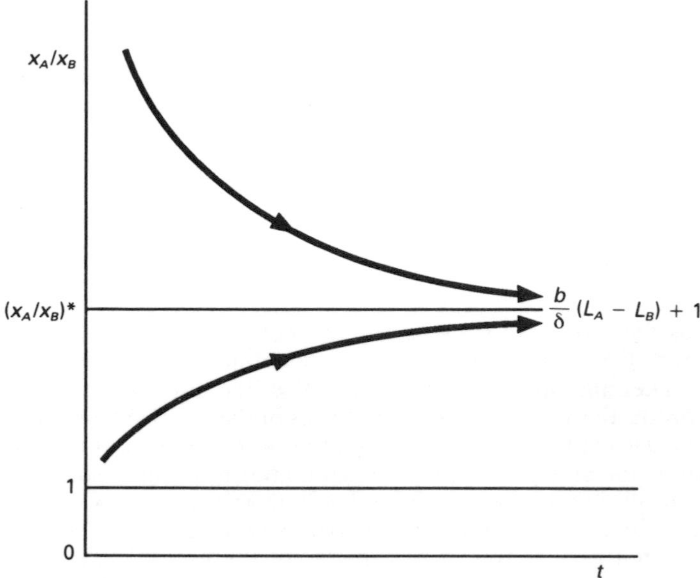

Figure 7.1 Convergence pattern of x_A/x_B

However, what if x_A/x_B is large enough to begin with, in particular, x_A/x_B exceeds this critical ratio? The logic of equation (7.9) entails that, with such initial values of x_A/x_B, the growth advantage that A has from its larger allocation of labour to the knowledge industry is overpowered by the diffusion effect and the diminishing returns effect. The advantage in growth then shifts from A to B and this results in reducing the size of the gap between A and B in the stock of knowledge and, correspondingly, in productivity levels. In this case, the gap asymptotically converges from above to the critical ratio $(x_A/x_B)^*$.

Thus, no matter how small or large the initial gap in the stock of knowledge between leader and follower, this process operates to bring about convergence in terms of growth rates of the stock of knowledge. But, as to the size of the gap itself, there is a sharp asymmetry, as shown in Figure 7.1. For small initial gaps, the size of the gap widens, up to some upper limit. For large enough gaps to begin with (that is, gaps larger than the critical ratio), there is a reduction in the size of the gap. However, whatever the initial size of the gap, there always remains a gap and it is positive. This result follows from the fact that the critical ratio is necessarily greater than 1, given that $L_A > L_B$, $b > 0$, $\delta > 0$. The magnitude of this permanent gap is uniquely determined by the difference in the allocation of labour to the knowledge industry, $L_A - L_B$, by the marginal product of experience, b, and by the diffusion parameter, δ.

Case 4: $1 > \delta > 0$, $L_B < L^* < L_A$

If country B is unable to achieve the threshold size of allocation of labour to the knowledge industry, then it is unable to participate actively in the productivity race. It remains in a dependent status of receiving whatever spillovers it can get from those already in the race, and its productivity level continues to fall further and further behind relative to the rest.

CONCLUSION

So far as the process of convergence/divergence in productivity growth is concerned, the analysis presented here identifies exactly what form that process takes and the conditions which affect the outcome.

The productivity gap is analysed in terms of the relative size of the stocks of knowledge existing in the leader and follower countries. It is shown that, given some initial gap to begin with (no matter how big or small), the gap asymptotically approaches a definite size from above or below depending on initial conditions. Whether the gap increases or diminishes depends on how big the initial gap is. Thus it is a matter of the exact degree of 'relative backwardness' in a precise sense, specified in relation to the critical ratio $(x_A/x_B)^*$. In particular, only if the initial gap is 'large enough' does convergence occur. In this respect, this result serves to give a certain precision to the well-known hypothesis of relative backwardness as a factor determining the tendency to convergence in productivity levels.[17] This result also replicates the diversity of the empirical record, insofar as that record exhibits the coexistence of dual tendencies of convergence and divergence. The coexistence of these two tendencies is shown here to be precisely connected with the cross-country distribution of initial conditions and parameter values around the critical ratio.

The analysis supports the need to maintain a sharp distinction between convergence in growth rates and convergence in terms of levels. In one class of cases, depending on initial conditions, even though growth rates converge, levels diverge and the gap correspondingly widens (albeit to an upper limit).

It is evident also that the process of productivity growth, under the conditions specified here, operates to keep the size of the gap within bounds. This is for reasons related, first, to the existence of a 'maturity' effect in the leading country associated with diminishing returns to experience. Second, it is related to the advantage that the follower country gains from the operation of a diffusion effect, or of 'spillovers' from the leader.

If the gap does not explode, neither is it ever eliminated altogether. A certain positive size of the gap is permanently reproduced by this process. That size, given by the critical ratio $(x_A/x_B)^*$, is uniquely determined by specific conditions of the productivity-increasing process: namely, the marginal product of experience, b, the diffusion parameter, δ, and the difference in relative allocations of labour to the knowledge industry, $L_A - L_B$.

A special class of cases consists of those countries that are unable to mount the scale required to start up the productivity-increasing process. In such cases the gap increases without limit. The same result would occur if the diffusion parameter were negative, implying that there are retarding effects or negative spillovers from diffusion.

One can readily admit that this analysis does not, and neither does it attempt to, tell the whole story concerning the historical record sketched in the first part of this chapter. It does provide a heuristic framework with which to identify various essential elements of the story that need to be explored in greater depth in seeking to explain the record of productivity growth.

As it stands, the model focuses on the character of the convergence/divergence process that occurs over a period of time appropriate to what one might call 'a given technological paradigm', during which it might be said that the frontier of technological knowledge is relatively fixed. It is in that context that it would seem to make sense to talk about 'diminishing returns to experience'. But, over the long haul, the paradigm does change and the frontier shifts along with it. This introduces the possibility that, by leapfrogging, followers may overtake and surpass leaders, so that the pattern of leadership changes. It would remain to determine who leads and who follows under those conditions, and whether there is any tendency to convergence. This effect is not considered here and is intrinsically more difficult to model.[18]

Another effect not captured here, which may be considered a significant part of the empirical record of productivity growth, is the intersectoral effect associated, for instance, with a shift from agriculture to manufacturing industry, or from traditional manufacture to services. This aspect of the process is essentially eliminated at the highly aggregative level of this model. For the same reason, it is not possible to capture a significant dimension of the process that is related to the effect of commodity specialization among countries.

Notes

1. A first draft of this chapter was circulated in February 1985. It was subsequently presented to the ASSA (Allied Social Science Association) meetings in New Orleans in December 1986, at various invited seminars, and at the conference from which this volume is derived. I wish to thank the editor of this volume, Ross Thomson, for his careful reading of the penultimate draft and for substantive comments and discussion of various points.
2. See for instance, Abramovitz (1986), Baumol (1986), Maddison (1982, 1987) and Matthews, Feinstein and Odling-Smee (1982).
3. See Baumol and Wolff (1988), De Long (1988), Romer (1989), Dowrick and Nguyen (1989) and Baumol, Blackman and Wolff (1989).
4. See Harris (1986), Baumol and Wolff (1988), De Long (1988) and Romer (1989).
5. For a penetrating discussion of some of the complex issues involved in analysing

the historical record of differential productivity growth among countries, see Abramovitz (1986) and Nelson (1981).

6. For reviews of the analytical elements and empirical studies in this burgeoning field of research, see Nelson (1981), Dosi (1988), Freeman (1982), Kamien and Schwartz (1982) and Stoneman (1983). For a related effort at modelling these effects in a growth-theoretic context, see Romer (1986, 1989).
7. The most forceful statement of the arguments supporting this view of the role of the knowledge industry, as a general perspective on the development of modern technology in the twentieth century, is that of Freeman (1982).
8. Fritz Machlup (1962) gives a much wider definition of the 'knowledge industries', and estimates that 30 per cent of the labour force in the US economy is included in his definition. Porat (1977) defines a similar category of 'information occupations' to include about 50 per cent of total occupations. I follow Freeman (1982, p. 5) in conceiving of what he calls 'the Research and Development system' as 'the heart of the whole complex'. Even so, there are still considerable empirical difficulties in identifying exactly what constitutes this 'system', as shown in National Science Foundation (1987).
9. For relevant empirical evidence on the contribution of R&D to productivity growth, see Griliches (1986) and Mansfield (1980). From the standpoint of ascribing causality, the relationships involved are considerably complex, as argued forcefully by Nelson (1981).
10. See Arrow (1962b) and Rosenberg (1982, Ch. 6).
11. Some relevant issues concerning the conception of knowledge as a commodity with its own peculiar attributes are considered by Arrow (1962a) and Nelson (1959).
12. Ames and Rosenberg (1963) review and dissect some of the arguments that have been put forward to support the broad thesis of an eventual retardation of growth in advanced industrial economies. They conclude that there is no logical necessity for such retardation and that the outcome depends on a number of empirical conditions that remain to be verified.
13. The logic of this idea does not necessarily rule out the existence of an initial phase of increasing returns, but for simplicity this analysis focuses on the case of diminishing returns as an 'ultimate' phase of the process. It must be emphasized that, as presented here, there is nothing inherent in the idea of 'diminishing returns to experience' that makes it a purely technological condition; rather, it is considered to be an analytic expression for a wide range of social and institutional factors that are themselves the product of historical development.
14. Issues involved in determining the pattern of trade and specialization among countries are left out of this analysis, inasmuch as the pricing side of the picture (on which an account of comparative advantage must be based) is being ignored. Still, it is not unreasonable to suppose that both countries produce some of the same two commodities, unless it should turn out that specialization according to comparative advantage yields a corner solution, which would be a very special case.
15. This assumption could be usefully dropped and the model extended to allow for differences in production conditions across countries.
16. The latter case is an implication of the argument that relations between advanced and underdeveloped economies are characterized by a 'structure of dependence' (see, for instance, dos Santos, 1970).
17. Abramovitz (1986) provides a discussion of this hypothesis, along with several extensions and qualifications, and reviews some of the historical evidence pertaining to it. The general idea that the 'degree of economic backwardness' is a

significant factor governing the pace and direction of development was put forward by Gerschenkron (1952).

18. For an examination of the issues involved in the question of overtaking and of changing leadership, see Ames and Rosenberg (1963).

References

Abramovitz, Moses (1986) 'Catching Up, Forging Ahead, and Falling Behind', *Journal of Economic History* 46, 2 (June).

Ames, Edward and Rosenberg, Nathan (1963) 'Changing Technological Leadership and Industrial Growth', *Economic Journal*, 72.

Arrow, Kenneth (1962a) 'Economic Welfare and the Allocation of Resources for Innovation', in *The Rate and Direction of Inventive Activity*, National Bureau of Economic Research (Princeton, NJ: Princeton University Press).

—— (1962b) 'The Implications of Learning by Doing', *Review of Economic Studies*, 29 (June).

Baumol, William J. (1986) 'Productivity Growth, Convergence, and Welfare: What the Long-Run Data Show', *American Economic Review* 76, 5 (December).

—— and Wolff, Edward N. (1988) 'Productivity Growth, Convergence, and Welfare: Reply', *American Economic Review*, 78, 5 (December).

—— Blackman, Sue A.B. and Wolff, Edward N. (1989) *Productivity and American Leadership: The Long View* (Cambridge, Mass.: MIT Press).

De Long, Bradford J. (1988) 'Productivity Growth, Convergence, and Welfare: Comment', *American Economic Review*, 78, 5 (December).

Dos Santos, Theotonio (1970) 'The Structure of Dependence', *American Economic Review*, 60, 2 (May).

Dosi, Giovanni (1984) *Technical Change and Industrial Transformation* (New York: St Martin's Press).

—— (1988) 'Sources, Procedures, and Microeconomic Effects of Innovation', *Journal of Economic Literature*, 26 (September).

Dowrick, Steve and Nguyen, Duc-Tho (1989) 'OECD Comparative Economic Growth 1950–85: Catch-Up and Convergence', *American Economic Review*, 79 (December).

Freeman, Christopher (1982) *The Economics of Industrial Innovation*, 2nd edn (Cambridge, Mass.: MIT Press).

Gerschenkron, Alexander (1952) 'Economic Backwardness in Historical Perspective', in Bert F. Hoselitz (ed.), *The Progress of Underdeveloped Areas* (University of Chicago Press).

Griliches, Zvi (1986) 'Productivity, R&D and Basic Research at the Firm Level in the 1970s', *American Economic Review*, March.

Harris, Donald J. (1986) 'Development of the World Economy: Convergence or Divergence?, paper presented to the National Economic Association Meeting, New Orleans, December.

Kamien, Morton I. and Schwartz, Nancy L. (1982) *Market Structure and Innovation* (Cambridge University Press).

Machlup, Fritz (1962) *The Production and Distribution of Knowledge in the United States* (Princeton, NJ: Princeton University Press).

Maddison, Angus (1982) *Phases of Capitalist Development* (New York: Oxford University Press).

—— (1987) 'Growth and Slowdown in Advanced Capitalist Economies: Techniques of Quantitative Assessment', *Journal of Economic Literature*, 25, 2 (June).

Mansfield, Edwin (1980) 'Basic Research and Productivity Increase in Manufacturing', *American Economic Review*, December.

Matthews, Robin C.O., Feinstein, Charles H. and Odling-Smee, John C. (1982) *British Economic Growth 1856–1973* (Oxford: Clarendon Press).

National Science Foundation (1987) *Science Indicators* (Washington, DC).

Nelson, Richard (1959) 'The Simple Economics of Basic Scientific Research', *Journal of Political Economy* (June).

—— (1981) 'Research on Productivity Growth and Productivity Differences: Dead Ends and New Departures', *Journal of Economic Literature*, 19 (September).

Olson, Mancur (1982). *The Rise and Decline of Nations: Economic Growth, Stagflation, and Social Rigidities* (New Haven, Conn.: Yale University Press).

Porat, M.U. (1977) *The Information Economy: Definition and Measurement*, vols 1–9 (Washington, DC: US Government Printing Office).

Romer, Paul M. (1986) 'Increasing Returns and Long Run Growth', *Journal of Political Economy*, 94.

—— (1989) 'Capital Accumulation in the Theory of Long-Run Growth', in Robert Barro (ed.), *Modern Macroeconomics* (Cambridge, Mass.: Harvard University Press).

Rosenberg, Nathan (1982) *Inside the Black Box: Technology and Economics* (Cambridge University Press).

Stoneman, Paul (1983) *The Economic Analysis of Technological Change* (Oxford University Press).

World Bank (1986) *World Development Report 1986* (New York: Oxford University Press).

Part III
Competition and Learning

The presence, rate and direction of technological change also depend on competitive factors. The relation between competition and technological change is complex. Technological improvement and diffusion are means to compete and are therefore encouraged by competitive pressures. Yet, when technological changes are costly, time-consuming or easily spread, competitive forces may discourage firms from innovating. By relating learning to competition, these chapters add a crucial dimension to our understanding of technical change in market economies.

Francesca Chiaromonte, Giovanni Dosi and Luigi Orsenigo model these interrelations of learning and competition. They begin by identifying three basic elements of an adequate model of technological change. It must incorporate invention and diffusion. It must allow for technological change to be embodied in capital goods, to alter the production of capital goods and to modify the process using the capital good. It must, finally, allow the costs of and the prospects for technological change to vary with the kind of technology, organization and norms of the firm. Chapter 8 then translates these requirements into a simulation model, which makes clear the critical roles played by path-dependent learning and competitive structure in the process of technological change.

Rejecting the common assumption that economic agents can acquire technological knowledge costlessly and instantaneously, Willi Semmler constructs a broadly evolutionary model with costly diffusion, competing techniques, imperfect information and increasing returns. His model combines three kinds of interfirm relationships that are usually modelled separately. Learning relations underlie diffusion among firms. Predator–prey relations spread new techniques by eliminating the old. Competitive relations can change market shares between old and new techniques in a variety of ways. He then constructs non-optimizing and optimizing simulations which, by showing that outcomes depend on the relative strengths of these factors, demonstrate the complexity of technological change.

In a comparison of Britain, the USA and Japan, William Lazonick further investigates the relation between institutional forms and innovation. Any successful industrial country must build the capacity to develop new techniques, but no two countries will follow the same institutional path. The way a country develops its innovative capabilities will affect its long-run technological and competitive success. In the long run, competitiveness requires firms to invest in the skills of its employees, while organizing to coordinate the knowledge of specialized workers. British firms failed to invest in their employees and thus could not compete with the rise of US managerial capitalism. US companies educated their managers but not their shop-floor workers, and thus have begun to lag behind the Japanese.

8 Innovative Learning and Institutions in the Process of Development: On the Microfoundation of Growth Regimes

Francesca Chiaromonte, Giovanni Dosi and Luigi Orsenigo

INTRODUCTION

In his recent assessment of growth theories, Moses Abramovitz refers back to the *Wealth of Nations* as the illustrious ancestor of a long stream of investigations on the determinants of economic growth (Abramovitz, 1989). In fact, in the famous proposition on the dynamics linking division of labour, productivity and market growth, Adam Smith identifies one of many positive feedback loops between innovative learning and economic development. Since then, the evidence on the microeconomics of learning and innovation has got much richer, especially in recent years (a survey in Dosi, 1988). However, standard growth theory has proved to be hardly suitable to incorporate the microevidence on, for example, dynamic increasing returns, path dependent learning, 'disequilibrium' search processes, interfirm and international differences in technological capabilities and so on, notwithstanding recent increasing returns equilibrium models (see Romer, 1986, 1990; Lucas, 1988; Aghion and Howitt, 1989; Grossman and Helpman, 1991). Neither has standard theory focused on the institutions and behavioural norms underlying economic coordination and development: that is, what are the institutional mechanisms that allow the 'Invisible Hand' to operate in a world that continuosly innovates? Again, that question can be traced back to Adam Smith, where he asks – in the *Wealth of Nations* and, especially, in the *Theory of Moral Sentiments* – what are the 'moral inclinations', beliefs and behaviours that make non-destructive interactions possible in market societies. However, since Adam Smith, attempts at an answer have mainly been left to disciplines other than economics. In that, the typical economic assumptions of 'perfect rationality' and equilibrium have often hindered any proper account of the sociology and 'political economy' of development.

To be fair, one has observed in recent years a renewed interest in the relationships between technical change, institutions and economic dynamics. For example, trying to interpret long-term historical discontinuities in growth patterns, as well as the diversity of these patterns across countries and over time, scholars like Freeman and Perez have introduced notions such as 'techno-economic paradigms' (or 'regimes'). This notion historically links the phases of sustained growth with the establishment of appropriate institutions and corporate behaviour governing technological learning.

Moreover, Boyer and the (mainly French) 'Regulation Approach' (see, for example, Boyer, 1988a, 1988b) have extensively argued that the major phases of economic development, as well as ruptures and crises, can be interpreted in terms of compatibility or 'mis-matching' among the prevalent behaviours of main social actors (firms, workers, social institutions and so on). In that view, technology, economic variables and institutions cannot be separately analysed. Rather, technical change is dynamically coupled with institutional change, and the growth, decline, stability or instability of various economies has ultimately to be ascribed to specific relationships between technological progress, institutional change (or lack of it) and economic signals.

Finally, a variety of contributions broadly in the Schumpeterian tradition has emphasized the crucial role in technical change and growth of decentralized and mistake-ridden processes of search, innovation and imitation, undertaken by agents that are heterogeneous in their capabilities, expectations and behaviour (Freeman, 1982; Nelson and Winter, 1982; Dosi and Orsenigo, 1988; Silverberg, 1987; Silverberg, Dosi and Orsenigo, 1988).

The model which follows tries precisely to provide a theoretical account – albeit a highly stylized one – of the growth process, fuelled by technological learning of heterogenous competing agents and embedded in particular institutional regimes. In the model, learning takes different forms, involving innovation and imitation in products (the search for new 'machines' that are more efficient for their users), in processes of production (the search for more efficient methods of production of the 'machines' themselves), and learning by using (on the side of adopters of those same 'machines'). The 'institutional regime' is captured by the type of norms which guide the behaviour of (highly 'boundedly rational') agents and by system-level parameters which characterize market interactions (on the products and labour markets, for example).

We start by assuming a wide set of notional opportunities of innovation. Such a set is permanently expanding via an endogenous process. Further, we model individual agents that are motivated in their innovative attempt by economic incentives, but also constrained and 'locked' by their competences. Finally, we represent an 'out of equilibrium' environment wherein:

1. individual agents, characterized by various and permanent forms of diversity, compete on the grounds of their specific technological achievements and behavioural rules (in each 'industry');

2. market interactions determine also intersectoral adjustments in demand, prices and, ultimately, the levels and changes of macroeconomic variables.

Within such a microfoundation of the process of coordination and change – broadly in an evolutionary perspective, as pioneered by Nelson and Winter (1982) – we shall explore:

1. the conditions under which microtechnological learning yields relatively ordered aggregate patterns of growth;
2. the effect of particular *norms* of behaviour and interaction among agents upon aggregate dynamics.

A central hypothesis in our approach is that the processes of technological innovation and those of institutional change (including the emergence, establishment and change in the norms of behaviour of the agents) *co-evolve*, although via different mechanisms and at different paces. Hence, one would ideally like a dynamic model whereby not only individual agents search in the technology space, innovate, imitate and compete with each other, but also endogenously learn about decision rules. Here, we fall short of this task. In a first approximation, we assume that behavioural norms, no matter how they have developed, present a relatively high degree of inertia *vis-à-vis* the quicker pace of technological learning and, thus, can be taken as parameters. So, in simulation exercises of 'comparative dynamics', we experiment with different (and fixed) distributions of the parameters describing decision rules, and compare the aggregate dynamics that they generate.

Developing and modelling some of the ideas discussed at greater length in Coricelli, Dosi and Orsenigo (1989), we shall show that the possible regularities in the dynamics of aggregate variables (such as income and productivity growth) are *emergent properties* of a system which self-organizes far from equilibrium. Moreover, our preliminary results show that different 'institutional regimes' do indeed affect both the rates of technological learning and the patterns of growth: only particular combinations of behavioural rules and market interactions appear to be viable to sustain macroeconomic growth.

In the following section we outline the basic hypotheses and the qualitative structure of the model. The next section – which the non-technical reader might want to skip – presents its formal description. The properties of the model are then explored via simulations and the results are discussed.

INNOVATION, COMPETITION AND GROWTH[1]

Ideally, a satisfactory model of growth should account for, or be consistent with, a series of 'stylized facts' on macrodynamics and on the microecon-

omics of innovation and competition. Several of these empirical regularities are discussed in Kaldor (1974), Simon (1984), Solow (1984), Maddison (1982). Greenwald and Stiglitz (1988) and Coricelli, Dosi and Orsenigo (1989). Let us briefly recall the stylized facts that are more directly relevant to the model which follows.
At the macro level:

1. output per unit of employment grows at relatively stable rates over rather long periods, and so do per capitum incomes;
2. there are persistent economic-wide business cycles;
3. one can identify 'phases' of development intertwined by 'crisis' and discontinuities in the average rates of growth of income and labour productivity;
4. the levels and rates of growth of income and labour productivity significantly differ across countries;
5. there is no systematic trend in capital/output ratios, and neither are there systematic correlations across countries between such ratios and the level of development;
6. significant levels of unemployment may persist for quite a long time.

Moreover, at the micro level:

7. innovation in products and processes is to a good extent endogenous to the activity of the business sector, via firms' R&D activities and also via more informal mechanisms of learning by doing and by using;
8. firms (and countries) systematically differ in their commitment to innovation and in their ability to innovate;
9. forms of market interactions different from pure competition are prevalent. Market structures are endogenous to the process of innovation and competition, with a lot of variability in market shares, relative costs and relative profitabilities of individual firms.

The theoretical task is to model a process of innovation, competition and growth whereby these 'stylized facts' can in principle hold together. The model presented here consists of two vertically connected sectors. Sector 1 produces heterogeneous (durable) production inputs ('machines') with labour alone. Sector 2 produces a homogeneous consumption good with labour and 'machines'. The bi-sectoral structure is clearly more apt than one-sector models to capture the transmission of demand and technological impulses between agents that might not have any competitive interaction with each other (that is, it may be taken as a metaphor for the input–output structure of the economy). Moreover, it makes easier the representation of both processes of endogenous innovation and diffusion. Labour is a homogeneous input. Each sector is composed of a finite number of firms. Their behaviour is represented as routinized, and they are characterized by a persistent variety in behavioural rules and technological capabilities. Firms' market shares endogenously change as a function of their relative 'competitiveness' (defined below), which, in turn, is connected with behavioural rules and technological capabilities. Firms of sector 1 innovate by developing new ('more efficient') types of machines and by finding more efficient techniques to produce them. Firms of sector 2 innovate by adopting and learning how to use the new types of machines. The collective outcomes of individual disequilibrium decisions determine also aggregate variables such as income growth, employment, wages and so on.

Innovative Opportunities

The very possibility of innovation must rest on the permanent existence of some unexploited technological opportunities, or the continuous emergence of new ones. In line with a growing body of evidence from the microeconomics of innovation (Rosenberg, 1976, 1982; Freeman, 1982; Dosi, 1988), we assume that unexploited opportunities permanently exist and that what is actually explored is much smaller than the set of notional possibilities. Furthermore, we draw a fundamental distinction between *knowledge* and *information*. Innovation, we suggest, does not imply only some processes of information acquisition about new products or techniques, but is grounded in pre-existing, partly tacit, forms of problem-solving knowledge embodied in the agents themselves. Innovation, in turn, augments this knowledge for individual agents or collections of them.

All this implies that, unlike the representation of technical change in the conventional growth or business cycle models, agents are not constrained by 'nature' in their innovative achievements, but rather by their own capabilities. More emphatically, paraphrasing Milton Friedman, there are always (semi-) 'free lunches', whose exploitation is primarily limited by agents' abilities.[2]

Our model of innovation incorporates the following major characteristics:

1. innovation involves changes in both products and production processes;
2. opportunities to innovate expand over time as a function of income growth (plausibly carrying with it more basic research, more scientific advancements and so on);
3. technical change is of two types, 'incremental' and 'radical';
4. abilities to innovate and imitate are firm-specific and depend on their past innovative record (that is, innovative learning is cumulative);
5. although some of the economic benefits from innovation and adoption of new products and processes can be appropriated by the innovators themselves, there are learning externalities: the ease of imitation varies with the number of incumbents already producing a certain commodity, and the skills in using a particular type of equipment partly 'leaks' from individual adopters to the whole industry.

In the model, as depicted in Figure 8.1, innovative opportunities in sector 1 correspond to 'machines', x, each of which corresponds to a technique of production for the final good, characterized by the labour productivity

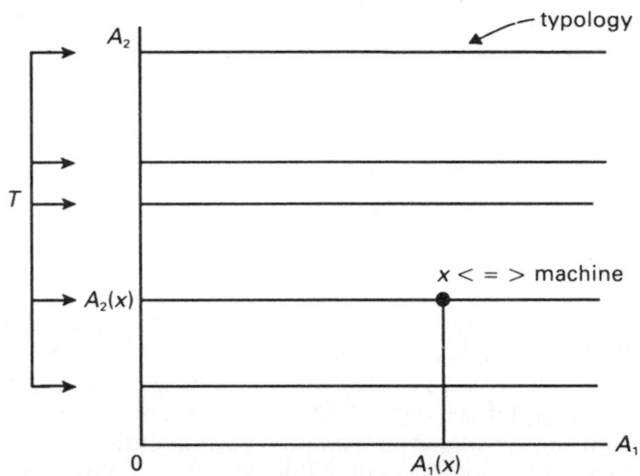

Figure 8.1 Innovative opportunities

coefficient $A_1(x)$, and to a technique of production for the machine itself, characterized by the labour productivity coefficient, $A_2(x)$. (Machines are 'measured' in terms of productive capacity, so that the machine input coefficient per unit of output is equal to one for each technique in sector 2 for each machine.)

In analogy with several empirical studies on innovation pointing out the difference between 'incremental' change – on the grounds of an unchanged knowledge base (a given 'paradigm') – and a 'radical' change (a new technological 'paradigm': see Dosi, 1984), we assume that the former is represented by increments in A_1, given A_2, while changes in both A_1 and A_2 correspond to 'qualitative' changes in technological competences and production organization (radical innovations). Hence, although empirically implausible, we assume for modelling simplicity that (within a paradigm) incremental innovation is only product-innovation for sector 1 agents (that is, it is the discovery of new machines embodying more efficient techniques for the final good production process – higher value of $A_1(x)$ – but produced with the same technique, a constant value of $A_2(x)$).

The 'notional opportunities', $T(\cdot)$, are defined by the set of achievable typologies (set of values for A_2) at each time $(\cdot)$. Over time, this set (containing old and new *potential* technological paradigms) grows via a stochastic process dependent on the level of economic (and scientific) development.

Note that the opportunity set always grows, or at least remains the same. However, as we shall see, the dynamics in potential opportunities does not directly influence the actual processes of innovation undertaken by individual agents. Agents' effective explorations generally remain well below the maximum notional opportunities.

Each agent, i ($i = 1, 2 \ldots n$), of sector 1 actually 'explores' the set of notional achievable typologies, $T(\cdot)$, via an expensive process based on R&D investments, assumed to be measured by the number of R&D workers that it employs. Such a search is represented as a stochastic process: when the search turns out to be successful the newly discovered machine will be a realization ($xg_i(t)$) of a probability distribution dependent on the production competences of the agents (as represented by the machine that is currently manufactured, $x_i(t)$), and on the opportunity set $T(t)$. Incremental progress within a typology of machines (elsewhere defined as the progress within a 'technological paradigm', or along a 'technological trajectory': see Dosi, 1984) is always possible, but subject to a progressive exhaustion of opportunities. This is like saying that any trajectory is characterized, in stochastic terms, by dynamic increasing returns, but at decreasing rates.

Conversely, the more a firm proceeds along a 'trajectory', the more its probability increases to jump on another one characterized by a higher value of A_2 (that is, to 'discover' a radical innovation). In turn, the discovery of a new 'paradigm' reopens the opportunities for further incremental improvements. Underlying this formalization, there is the idea that the progressive exhaustion of a trajectory is a 'focusing device' (Rosenberg, 1982) which triggers the search for new sources of innovative opportunities, while past experience on a trajectory also cumulatively built technological competence for the discovery of new trajectories.

The exploration process as a whole maintains its increasing returns and cumulative nature, with periods of incremental technical progress intertwined with more discrete jumps in both product and process technologies.

Moreover, technological knowledge, in most circumstances, is neither a purely public good nor an entirely appropriable blueprint or patent (such as in the formalization of 'patent races'), but rather can be reproduced and imitated in an expensive way (Mansfield, Schwartz and Wagner, 1981; Pavitt, 1987; Dosi, 1988). In the model we introduced the possibility, for each agent of sector 1, of imitating machines that are known, although not necessarily produced, by other agents.

We formalize imitation as a stochastic process whose success depends on the amount of imitative R&D investment undertaken by each firm. When such a search is successful, we further assume that, first, the probability of imitating any one type of machine is inversely related to the technological distance from the current capabilities of the imitating agent (given the partly tacit nature of technological knowledge, it is easier to imitate machines which are relatively similar to those that the agent already produces or knows). Second, the more widespread is the knowledge about a particular machine, the easier might be innovation: thus, we assume that the probability of imitating any one type of machine is directly proportional to the absolute frequency of agents who know it.

Through the search activity of sector 1 agents, the system continuously 'discovers' (and produces, under microeconomic decision rules specified further on) new capital-embodied opportunities for sector 2.

Innovation, for each agent j ($j = 1, 2 \ldots m$), of sector 2 is equivalent to the purchase of new types of machines, via expansion and scrapping decisions (whose rules, again, shall be specified below). However, producers of the (homogeneous) final good must learn how to use each machine type (A_2) up to the optimal use defined by the value of A_1 of the machines belonging to that type (that is, belonging to any one 'technological paradigm'). Similar to Silverberg, Dosi and Orsenigo (1988), we assume that each j embodies a *skill*, $s_j(A_2, t)$ (a coefficient between 0 and 1) in the use of machines belonging to the typology A_2, so that the actual labour productivity it can achieve on them at time t might well be less than the 'optimal' value. These skills improve via two learning processes. The first is a learning-by-using activity (as in Rosenberg, 1982), whereby the capabilities of using each typology of machines grow with the actual cumulated use of them. We assume that *private* firm-specific and technology-specific learning follow a logistic path. However, we assume also a second process of learning, through which part of the private skills 'leak out' to other potential adopters, via information diffusion, personnel mobility, consulting activities and so on (Silverberg, Dosi and Orsenigo, 1988); this part becomes a *public* skill shared by every firm.

Therefore, late-adopters of a particular typology A_2 in t will start by using only public skills, which in turn depend on previous adoption: that is, on the number of adoptions, the opportunities of private learning to use, and the appropriability of private capabilities (inversely related to the rate at which private skills become public).

To summarize: in the model, the economic system continuously explores new opportunities of capital-embodied technical progress, via R&D-based processes of generation and imitation of 'machines', undertaken by agents in sector 1. Moreover, the flow of new machines represents also a continuous opportunity of innovation for the agents in sector 2, via their adoption, the subsequent process of learning how to use them, and the learning externality that industry-wide diffusion of information and skills provides for late-adopters. Finally, the aggregate income dynamics, specified below, induces an endogenous increment of new potential avenues of innovation. All this defines an expanding universe of *notional* and (partly) *exploited* opportunities of technical progress.

However, in a microfounded model, activities of search, innovation, imitation and adoption by profit-motivated agents must be grounded in some economic incentive to do so and, more generally, in microeconomic decisions. This is what we shall discuss in the following.

Microeconomic Decision Rules

It is a standard practice in the economic profession to assume perfectly optimizing agents (and growth theory is no exception). It is impossible to discuss here the analytical status of that assumption. Suffice to say that such a representation of the decision processes, irrespective of its empirical plausibility under stationary conditions (that is, given technologies and preferences) becomes particularly demanding on the computational and forecasting capabilities of agents when innovations may occur endogenously. As Herbert Simon put it:

> [i]n a formal way, it is perfectly feasible to produce a theory of technical innovation based on the postulate of rationality. Since innovation is presumably produced by investment of human and capital resources, we introduce a new production function for innovation, and equate the value of marginal product of innovation with its cost. From a formal standpoint, we have simply replaced the task of estimating the parameters of a function, the production function for innovation. The only obvious gain from the replacement is that we can now rest comfortable in the knowledge that everything is proceeding rationally. Human rationality is now only bounded by the characteristics of the external environment: the quality of the ore that is mined by the innovation production process. Of course, if we examine the metaphor too closely, we see that 'quality' of the 'ore' is an euphemism for the 'effectiveness of the thought processes of human beings who are doing the innovating'. (Simon, 1984, pp. 41–2)

Indeed, if one sticks to a 'rational' microfoundation of economic dynamics *cum* endogenous innovation, one must subscribe to either (or both) of the following hypotheses, namely, (a) the actual 'thought processes' rather closely resemble optimizing procedures (and generally involve also 'rational' technological expectations), and, (b) if and when hypothesis (a) does not hold, the competitive process is such that the relative frequency of behaviours that *ex post* turn out to be 'optimizing' rapidly tends to 1.

In fact, good empirical and theoretical reasons – related to the nature of the innovation process, the associated uncertainty, the characteristics of innovation-based competition and so on – strongly suggest that hypothesis (a) does not generally hold (much more on that in Nelson and Winter, 1982; Simon, 1984; Winter, 1987; Dosi and Orsenigo, 1988; Dosi and Egidi, 1990). And, regarding hypothesis (b), its proponents have so far failed to deliver any proof of the general stability of rationally microfounded equilibrium dynamics whenever perturbed by 'disequilibrium behaviour'.

Here, we want to explore an opposite behavioural foundation. We assume that agents are characterized by *fixed decision rules* – generally different in their parameters among agents – concerning investments, research, pricing and so on.

In principle, these 'rules' must have developed via some *learning processes* over the history of individual firms and of the system as a whole. Here, however, we do not model such learning processes. We set ourselves the task of showing that some orderly aggregate properties may emerge even in those extreme conditions when *no* individual behaviour can be presumed to be an 'equilibrium' (optimizing) one, and also when no form of decision learning is going on. After all, the common modelling assumption of perfect rationality derives a good deal of the purported system properties from the forecasting and computing abilities of the 'representative' agent(s).

Moreover, whenever one adopts such a compact microfoundation (a single agent representing the whole of them), one also assumes by hypotheses perfect microeconomic coordination and precludes the possibility of analysing the long-term effects of those non-average behaviours that are typically associated with innovative entrepreneurship. In this respect, our agents are much less able to forecast and compute 'equilibrium behaviours' than commonly assumed, but they are continuously prepared to (imperfectly) adjust to the unexpected, and able to generate and imitate novelty. They are *different* in their decision parameters, although not in the rules themselves. Finally, they differ in their technological abilities and degrees of success in innovation and imitation, in line with the general finding from the innovation literature that technology presents highly firm-specific, tacit and idiosyncratic features.[3] In that, we entirely share and push further Greenwald and Stiglitz's suggestion that:

> [i]ndividuals do not have perfect foresight or rational expectations concerning the future. The events which they confront often appear to be unique, and there is no way that they can form a statistical model predicting the probability distributions of the outcomes. There is little evidence that they even attempt to do so. At the same time, individuals are not myopic. They do not simply assume that the future is like the present. (Greenwald and Stiglitz, 1987, p. 131)

Our decision rules implicitly embody also firm-specific 'theories' on how to cope with a changing world. This implies also that there is no common knowledge of either an 'innovation production function' or even a unique 'production function': both differ across firms and change over time according to agent-specific patterns.

Decision Rules: Sector 1

Machine producers do know that notional opportunities for 'better machines' always exist, and they know also the general procedure of discovering them (that is, by investing in R&D). Finally, they expect to derive some economic benefits from successful innovation or imitation. New machines are partly appropriated, due to various form of knowledge specificity (with or without a patent system), and indeed it turns out that some innovators increase their market shares and their profits. Similarly, successful imitation yields a form of 'business stealing'.[4] The fact that, *for some*, innovation and imitation result in differential profits and increasing market share, implies what we call *weak incentive compatibility* of search activities. However, agents do not know the actual probabilities of innovation or imitation, or the precise technical features of what they will discover. Furthermore, the product-market sales and profits of each firm depend on the performances and prices of every machine produced by all firms, which in turn are a function of the success in innovation and imitation by each of them. In such circumstances we make the extreme assumption that agents determine the level of their decision variables via fixed and firm-specific rules.

In sector 1, each firm is characterized by the following decision rules:

1. *desired* allocation of resources to R&D, via a fixed percentage of previous period turnover (note that here and throughout, individual agents may be unable to achieve their 'desired' rates of investment and production, given their maximum available liquidity determined by cash flow plus bank credit, the latter being a fixed proportion of the former);
2. a parameter distributing R&D between innovative and imitative search;
3. desired prices on each new machine that is or could be manufactured, via a mark-up on direct *costs* of production;
4. an evaluation function comparing prices and performances of currently produced machines with those stemming from innovation and imitation if any, in order to decide which one to manufacture (we assume single-product agents);
5. desired level of production for whatever machine is chosen.

Sticking to our general endeavour of exploring the possible orderly characteristics of system coordination and change in presence of highly routinized microbehaviours, we assume a very simple adaptive expectation which extrapolates from the demand of firm i over the previous two periods.

Given these five decision rules the desired demand for labour can also be automatically derived.

Decision Rules: Sector 2

The industry producing the final good is somewhat similar to what Pavitt (1984) calls a 'supply-dominated' sector: innovations are generated elsewhere in the system and introduced via decisions of scrapping and/or expansion. The levels of production and scrapping, jointly with the choice of the type of machine to purchase, influence also production efficiency, and thus unit costs of production.

Firms are characterized by the following decision rules.

1. *Desired level of production*. As in sector 2, it is derived from adaptive expectations on demand, based on their own demands of the previous two periods. (Given the existing capital stock, the rule allows also the automatic determination of desired net investment.)
2. *Desired price*, again, via a mark-up over direct unit costs.
3. A *choice function* which compares prices and performances of the new machines to purchase, if any.
4. A *scrapping rule*, based on a pay-back period criterion.

As in section 1, these rules are sufficient to determine also the desired demand for labour.

Competition, Coordination and Aggregate Dynamics

Firm-specific decision rules, together with the history of innovation, imitation and learning by individual agents, account for a permanent diversity among them. Firms differ in (a) the quality of products they offer (sector 1); (b) the prices they charge (both sectors); and (c) the parameters of the decision rules concerning their control variables (again in both sectors). All this, coupled with the fact that no agent is capable of correctly computing *ex ante* equilibria where individual actions are reciprocally consistent, implies that competition is a *permanent disequilibrium process*. Its outcomes are also variations in the market shares and profits of individual firms. Hence markets perform as *selection mechanisms*, acting upon the relative competitiveness of individual actors. In the model, as in common language, competitiveness is a relative concept: it implies a comparison among different agents. In sector 1, the competitiveness of each firm i is determined by (1) the production efficiency for the users of the machine offered on the market (as expressed by the value of $A_1(x)$); (b) its price (connected to the value of $A_2(x)$); and (c) the degree to which the firm might 'ration' the market since it might not be able to fulfil all demand given forecasting errors and/or liquidity constraints. Similarly, in sector 2, the competitiveness of each firm j is determined by (a) the price of the output (note that, despite the homogeneous good, we allow for prices that are different among firms); and (b) the possible degree of 'rationing' exercised by the firm (as described above for sector 1).

Firms with above-average competitiveness expand in their market shares, while firms below average shrink or even die. However, the selection criteria themselves are endogenous: they are the collective outcome of the dynamics of the competitiveness of each firm. Consider the selection process in sector 1, which is endogenous in two ways. First, it is determined by the individual dynamics of product innovation (that is, the changes in A_1 – and in the typologies A_2 – achieved by each i), in prices and in production capacity. Second, it is also determined by the evolution of the users' sector, via learning by using and expectations of machines users. Similarly, selection in sector 2 is shaped by the individual dynamics of capital-embodied process innovation, learning by using and the externality that all this entails.

In a biological analogy, the selection processes of the model represent the equivalent of a 'fitness function'. However, the 'landscape' over which 'fitness' is defined *co-evolves* with the behaviours and technological characteristics of individual agents in both sectors. In such an environment, individual abilities of forecasting the technology and the behaviours of the others is highly imperfect. Each agent can be expected to face unfulfilled expectations and lack of correspondence between *ex ante* plans and *ex post* realization. Therefore the agents must possess some adjustment rule in their control variables. In particular, with regard to prices, we model an (imperfect) adjustment rule whereby each firm changes *actual* prices while trying to strike some balance between desired (mark-up) prices and its relative competitiveness on the market (as revealed by the dynamics of market shares).

The system coordinates and evolves via competition among heterogeneous agents which continuously introduce, adopt and imitate technological innovations. In turn, the outcome of the competitive process feeds back to individual behaviours both directly, via price adjustments, and indirectly, via the effects that relative competitiveness exert on firm size and cash flow (and thus, also, on investments in R&D and machines).

Aggregate ('macro') variables are the endogenous outcomes of such processes of innovation and competition. Aggregate gross and net investment in machines sum, *ex post*, over the investments of individual firms. So does employment.

In line with the unorthodox spirit of this work, we further add the following assumptions: first, that labour supply is unlimited at any positive wage, and wages change, in the most general formulation, as a function of (a) previous period variations in labour demand; (b) changes in average system-level labour productivity; and (c) changes in the final good prices. Indeed, we shall represent different *labour market regimes* by allowing varying combinations of 'institutional' mechanisms of indexation of wages on cost-of-living and productivity of labour, and a 'classical' adjustment on net variations in labour demand.

Second, in an extreme and out-of-fashion 'Keynesian' simplification,[5] all wages are consumed (final good), profits are either invested (R&D or machines) or saved in the form of interest-yielding deposits; the interest rate is exogenously fixed, and does not have any effect on investment decisions. There is no credit ceiling at a system level.[6]

Indeed, these extreme assumptions, together with the earlier ones and the nature of technological change and rule-driven behaviour, define a system that is highly *unconstrained*. It is not bounded by 'nature' in its dynamics; unlike standard models, one cannot ultimately find a principle of order in the scarcities of primary inputs jointly with some objective technological constraint. Moreover, highly institutional (rule-driven) behaviour cannot provide by itself those ordering virtues which a perfectly rational 'representative agent' with foresight is assumed to possess.

These features of our model are instrumental in investigating the conditions under which a system that continuously 'explores' a wide and expanding set of possible states can self-organize despite, or *because of*, disequilibrium microbehaviour. Ideally, the non-linear dynamics of the system should be able to generate the 'stylized facts' recalled at the beginning of this section: here, we shall simply present some highly preliminary results. However, before doing that, we shall present more formally the structure of the model.

A SELF-ORGANIZATION MODEL

Innovation, imitation and Production: Sector 1

Machines are identified by couples of integers (A_1, A_2), corresponding to points in the non-negative subset of a bidimensional space $(\mathbb{R}^2_+)$.[7]

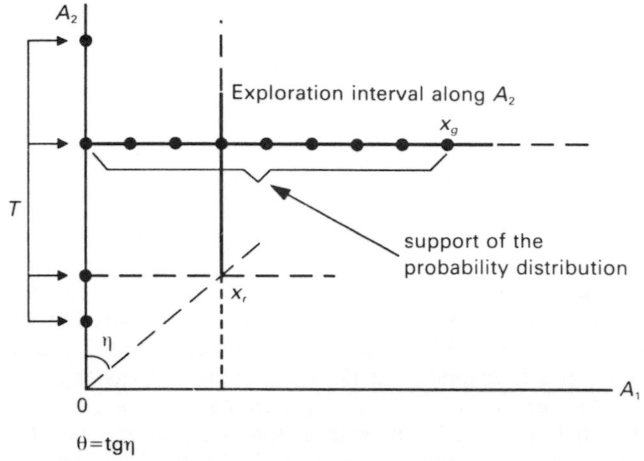

Figure 8.2 Innovation in sector 1

We define each 'typology' of machines belonging to the same paradigm as a set of points, $x = (A_1, A_2)$ in $\mathbb{R}^2_+$, given by

$$(x \in \mathbb{R}^2_+ : A_1 \in \mathbb{N}, \text{ and } A_2 = a_2), a_2 \in \mathbb{N}$$

in correspondence of various values of a_2 (see Figure 8.2). The set of achievable typologies, $T(\cdot)$, grows via a 'two-stage' stochastic process, the first stage determining the access to the second (see Nelson and Winter, 1982). In the first stage, the stochastic event 'access to the creation of a new paradigm' is determined as a draw from a Bernoulli distribution with probability

$$Pr\ (AT(t) = 1) = 1 - \exp\ (- r \cdot Y_m(t)) \tag{8.1}$$

where $Y_m(t)$ represents a moving average of aggregate income up to time t.

If access occurs, the system adds a new typology to the set, T, of the achievable ones, via an equiprobability draw in the set of the integers within the closed interval

$$[A_2^*(t-1); (1+h) \cdot A_2^*(t-1)]$$

where $A_2^*(t-1)$ is the maximum value in the set $T(t-1)$. r in equation (8.1) and h are parameters that implicitly capture the general state of '*scientific*' opportunities and the effectiveness of their exploitation (translation into achievable paradigms). In order actually to access new 'machines', each agent, i, in sector 1 undertakes a two-step stochastic procedure, similar to the one above, through which it can access the generation of a new machine, in t, with probability

$$Pr\ (AG_i(t) = 1) = 1 - exp\ (-\ r_g \cdot I_i^g(t)) \tag{8.2}$$

where I_i^g represents i's investments in innovative search, in terms of number of R&D workers, and r_g is a system parameter. If the random access occurs, the new machine is drawn from a uniform probability distribution whose support is defined by the set of points with second coordinate (A_2) given by

$$A_{2i}^g(t) = \max\ (T(t) \cap [A_2(x_i(t)),\ A_2\ (x_i(t)) + \lambda \cdot \theta_i(t)]) \tag{8.3}$$

and first coordinate given by the integers in

$$[\max\ (0;\ A_1(x_i(t)) - (1/\lambda) \cdot (A_{2i}^g(t)/A_1(x_i(t)));\\ A_1(x_i(t)) + (1/\lambda) \cdot (A_{2i}^g(t)/A_1(x_i(t)))] \tag{8.4}$$

where λ is a system parameter and

$$\theta_i(t) = A_1(x_i(t))/A_2(x_i(t))$$

This formulation implies that exploration along a 'trajectory' (holding A_2 constant) induces a relative shrinkage of further opportunities of incremental innovation, but also expands the set of attainable 'machines' on the other 'trajectories'. An illustration of this is given in Figure 8.2.

Agents can also imitate each other's machines (again, via a two stages stochastic process). The probability of access to imitation is

$$Pr\ (AM_i(t) = 1) = 1 - exp\ (-\ r_m \cdot I_{\mathrm{i}}^{\mathrm{m}*}(t)) \tag{8.5}$$

$I_{\mathrm{i}}^{\mathrm{m}}(t)$ is the investment in imitative search of agent i, and r_m is a system parameter which is lower the higher is the appropriability of innovations.

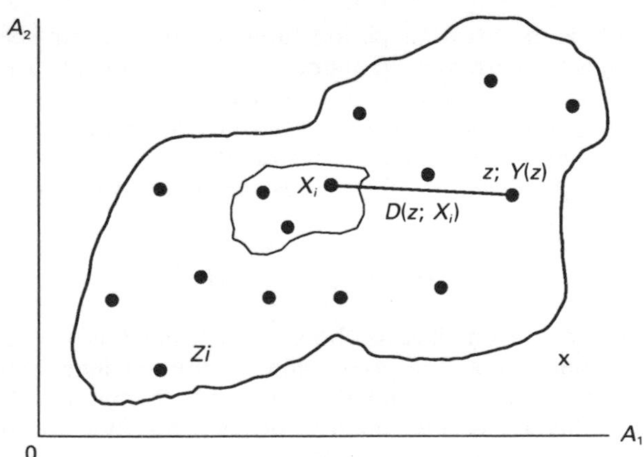

Figure 8.3 Imitation in sector 1

If access to imitation occurs, the imitated point (machine) will be a realization ($x_i^m(t)$) from a probability distribution defined over the set $Z_i(t) = X(t) \setminus X_i(t)$, where $X(t)$ is the set of the points known by the complex of the agents of sector 1, and $X_i(t)$ is the set of the points known by i (one rules out the possibility of imitating oneself!).

The probability of imitating any one type of machine in Z_i for i is inversely related to the distance, expressed in an orthogonal metric, between each point, $z \in Z_i(t)$, and the set $X_i(t)$.

$$D_{MET}(z, X_i(t)) = \min_{x \in X_i(t)} [(z - x)' \cdot \mathrm{MET} \cdot (z - x)]^{1/2}$$

where MET is the matrix 2×2

$$MET = \begin{pmatrix} met1 & 0 \\ 0 & met2 \end{pmatrix}$$

The probability is also directly related to the frequency of agents knowing any one kind of machine in Z_i, $\gamma(z, t)$. Thus, the probability of imitating any one machine, $z \in Z_i(t)$, normalizing, is given by

$$Pr(x_i^m(t) = z) = \frac{\gamma(z, t) \cdot [D_{MET}(z;X_i(t))]^{-1}}{\sum_{x \in Z_i(t)} \gamma(x, t) \cdot [D_{MET}(x;X_i(t)]^{-1}} \tag{8.6}$$

An illustration of this is presented in Figure 8.3.

Each firm, i, determines its total desired allocation in R&D according to the rule

$$I_i^d(t) = \mu_i \cdot S_i(t - 1) / w(t) \tag{8.7}$$

where $S_i(t - 1)$ is the previous period turnover, μ_i is a firm-specific parameter and $w(t)$ is the current wage (hence, R&D investments are defined in terms of number of research workers). Actual investments, $I_i(t)$, as well as the actual values of the other decision variables of the firm, correspond to the desired ones up to the liquidity constraint.

A firm-specific parameter, ζ_i, distributes research between innovative and imitative activities, so that

$$I_i^g(t) = \zeta_i \cdot I_i(t) \text{ and } I_i^m(t) = (1 - \zeta_i) \cdot I_i(t)$$

At each time, each agent has to choose which machine to manufacture between the variety already in production, x_i, the machine discovered via innovative R&D, if any, x_i^g, and that imitated, if any, x_i^m. That choice involves an evaluation of the 'quality' of each machine for the users – approximated by the 'optimal' cost of production on the machines, given current wages – and also of the supply price of the machines themselves (p_i):

$$V_i(z, t) = \phi_1 \cdot ln\,(w(t)/A_1(z)) + \phi_2 \cdot ln\,(p_i(z, t)) \tag{8.8}$$

V_i is the notation for the evaluation functional, and it has to be applied to $z = x_i; x_i^g; x_i^m$.

For 'new' machines (candidates coming from innovation and/or imitation, and never manufactured before by the agent), prices are *desired* prices, calculated via a mark-up rule,

$$p_i^d(z, t) = m_i \cdot (w(t)/A_2(z)) \tag{8.9}$$

(m_i being the firm-specific mark-up parameter), while for the machine already in production, the price is the *actual* price applied by the agent in the previous period (which, as we shall see, may well be different from the desired one). If all three candidates for production belong to the same typology of machines (that is $A_2(x_i) = A_2(x_i^g) = A_2(x_i^m)$) the choice is straightforward, and will satisfy a simple minimization of $V_i(\cdot)$. However, if one or both the new candidates belongs to typologies that differ from that previously manufactured ($A_2(x_i)$), their evaluation V_i is multiplied by a 'prudential' firm-specific coefficient $\chi_i/w(t)$. The prudential correction requires that, in order to switch production to a new paradigm, with the consequent market uncertainty and learning costs for the users, the expected superiority of the new machine *vis-à-vis* the previously manufactured one must exceed a certain threshold, influenced by the 'animal spirits' of individual entrepreneurs.

Having chosen which machine to produce (the new x_i for the current period), each agent defines its desired level of production in relation to its demand expectations and its carried over inventories. Expected demand is extrapolated as

$$D_i^e(t) = \delta_i\,(D_i\,(t-1), D_i(t-2)) \tag{8.10}$$

where δ_i is a firm-specific function expressing demand expectations derived from the level of actual demand in the previous periods.[8]

Thus, the level of desired production is

$$Q_i^d(t) = D_i^e(t) - N_i(t-1)$$

$N_i(t-1)$ being the inventories carried over from the previous periods of production. As for research investment, actual production level ($Q_i(t)$) is equal to the desired one, unless adjusted via the liquidity constraint. Production occurs with labour alone, under constant returns to scale, at unit costs $w(t)/A_2(x)$.

Diffusion, Learning and Production: Sector 2[9]

The machines are purchased and used by sector 2 producers of the final good. Each agent, j ($j = 1, 2 \ldots m$), utilizes each machine with a skill level $s_j(A_2, t)$ (bounded between 0 and 1) specific to the agent and to the machine type (A_2). Thus actual labour productivity on each machine x, such that $A_2(x) = A_2$, is

$$s_j(A_2, t) \cdot A_1(x) \leq A_1(x) \tag{8.11}$$

Let us define the set of typologies actually produced at each time (.) by sector 1 as $T^s(t)$. On the new typologies offered on the market for the first time in t (that is, on $T^s(t) \backslash T^s(t-1)$), the initial using skills are assumed to be identical for all potential adopters of sector 2, and equal to the system parameter, σ. An agent that adopts any one type, A_2, of machines will grow in its using skill according to

$$\begin{aligned} s_j(A_2, t) - s_j(A_2, t-1) = \alpha_1 \cdot [Q_j(A_2, t)/(CQ_j(A_2, t) + a_2)] \\ \cdot [s_j(A_2, t-1) \cdot (1 - s_j(A_2, t-1))] \end{aligned} \tag{8.12}$$

where Q stands for the production undertaken by j in t with machines of typology A_2, CQ is the correspondent cumulated production,

$$CQ_j(A_2, t) = \sum_{t=0 \ldots t} Q_j(A_2, t)$$

and α_1 is a system parameter capturing the level of learning by using opportunities.

Private skills, s_j, tend to partly diffuse to the whole industry. The *public* (commonly shared) machine using skill for each typology A_2 ($s_p(A_2, t)$), which starts at level σ when a typology is first introduced, grows over time according to

$$s_p(A_2, t) - s_p(A_2, t-1) = \beta_1 \cdot (s_m(A_2, t-1) - s_p(A_2, t-1)) \tag{8.13}$$

where s_m is the average of agents, j, private skills, weighted with their market shares, and β_1 is a parameter which is higher the lower is the private appropriability of machine using skills. Any new later-adopter of machines of a certain type (A_2) will start with public skill levels $s_p(.)$.

Each firm determines its levels of production based on a rough expectation rule on j's own demand

$$D_j^e(t) = \delta_j(D_j(t-1), D_j(t-2)) \tag{8.14}$$

(with a meaning analogous to that of equation (8.10) for sector 1), yielding desired production

$$Q_j^d(t) = D_j^e(t) - N_j(t)$$

N being, again, the inventories. The level of production allowed by the current capital stock is identical to the number of (heterogeneous) machines retained by the agent. Calling $\Gamma_j(t)$ the set of them (recall that they are measured in terms of units of production capacity), we have

$$K_j(t) = \sum_{x \in \Gamma_j(t)} g_j(x, t)$$

where $g_j(x, t)$ are the absolute frequencies of machines. The desired net investment (expansion) is

$$IE_j^d(t) = \max(0; \hat{K}_j^d(t))$$

$\hat{K}_j^d$ being the desired variation in capital stock, calculated as

$$\hat{K}_j^d(t) = \nu_j \cdot [Q_j^d(t) - u_j \cdot K_j(t)] \cdot (1/u_j) \tag{8.15}$$

where u_j is the firm-specific level of desired capacity utilization ($0 \leq u_j \leq 1$), and ν_j is a firm-specific parameter that can be taken to capture expectations about future levels of demand as compared to the past (note also that throughout the model own-demand expectations concern both industry-wide levels and one's own market share).

With a non-binding liquidity constraint, both actual production and actual expansion investment levels will equal the desired ones (otherwise, they are proportionally corrected).

Each j must choose within the set of machines currently offered by sector 1 producers ($X^s(t)$). It follows from the foregoing representation of technological learning that each agent, j, whenever choosing among different typologies, must account also for its specific skills in using them, $s_j(A_2, t)$. These skills imply a form of dynamic appropriability of learning by using. However, precisely because of such a partly appropriated learning opportunity, expectations of potential skill improvements enter the choice rule (like Silverberg, Dosi and Orsenigo, 1988). Call a_j the '*expectation bonus*'. The adjusted skills used for machines selection are

$$s_j^e(A_2, t) = s_j(A_2 t) + a_j \qquad \text{if } s_j(A_2, t) < 1 - a_j \tag{8.16}$$

$$= 1 \qquad \text{otherwise}$$

Moreover, as we shall see below, nothing guarantees that markets for each type of machines clear: the degrees of 'rationing' by each machines supplier, $l_i(x, t)$, negatively influence preferences.

The choice of the machine depends on its price, the labour productivity that it entails and its possible levels of rationing. The chosen machine will be the one, among the supplied ($x \in X^s(t)$), that maximizes the evaluation functional

$$M_j(x, t) = \phi_3 \cdot ln\,(s_j^e(A_2(x), t) \cdot A_1(x)) - \phi_4 \cdot ln\,(p_i(x, t)) - \phi_5 \cdot ln\,(l_i(x, t)) \quad (8.17)$$

Call $x_j^*(t)$ the chosen machine.

Note that more than one producer may manufacture the same machine, but possibly offer it at a different price. The choice of sector 2 agents is therefore among 'couples' (machines-producers) which identify both the machine characteristics (typology A_2 and 'quality' A_1), and the supply price and 'rationing' conditions (p and 1, respectively). Since machines rationing can occur, each j may be forced to 'second best' options.

As regards scrapping, we assume that it follows a pay-back period rule. The current capital stock $K_j(t)$ is made of different kinds of machines, x, associated with *actual* labour productivities $\pi_j(x, t) = s_j(A_2(x), t) \cdot A_1(x)$, and actual unit labour costs $c_j(x, t) = w(t) / \pi_j(x, t)$ ('optimal' unit labour costs given the machines stock are simply $w(t) / A_1(x)$).

Correspondingly, call $c_j^*(t)$ the actual unit labour cost associated with $x_j^*(t)$.[10] The subset of the capital stock that j desires to scrap will be

$$OS_j(t) = (x \in \Gamma_j(t): [p(x_j^*(t))/(w(t)/A_1(x)) - c_j^*(t)] \leq b_j) \quad (8.18)$$

where b_j is a firm-specific pay-back period parameter. Desired substitution investment will thus be

$$IS_j^d(t) = \sum_{x \in OS_j(t)} g_i(x, t)$$

Like producers of sector 1, also each j may be unable to meet its desired plans due to a liquidity constraint, in which case (in sector 2) we assume that firms will satisfy, in order, current production, expansion investments and replacement.

Production is characterized by constant returns to scale and the notional average productivity depends on the frequencies of different types of equipment

$$\pi_{mj}^n(t) = \sum_{x \in \Gamma_j(t)} (1/A_1(x)) \cdot (g_j(x, t)/K_j(t)) \quad (8.19)$$

However, actual average productivities, and average unit labour costs (c_m) depend on such frequencies, but also on typology-specific and firm-specific skills[11]

$$c_{mj}(t) = \sum_{x \in \Gamma_j(t)} (w(t)/\pi_j(x, t)) \cdot (g_j(x, t)/K_j(t)) \quad (8.20)$$

Finally, desired prices are determined via a mark-up over average labour unit costs, $p_j^d(t) = m_j \cdot c_{mj}(t)$.

Market Interactions and System Dynamics

Each sector 1 producer competes in the machine market for the demand stemming from expansion and replacement decisions of sector 2 manufacturers. Similarly, the latter compete on the final good market whose size is determined by the real wages of all employed workers.

In sector 1 the competitiveness of agent i ($E_i(t)$), manufacturing x, is defined as

$$E_i(t) = -\omega_1(t) \cdot ln\,(w(t)A_1(x)) - \omega_2(t) \cdot ln\,(p_i(x, t)) - \omega_3(t) \cdot ln\,(l_i(x, t-1)) \quad (8.21)$$

where $A_1(x)$ is the labour productivity of machine x for its users, $p_i(x, t)$ is the price charged for it by manufacturer i, and $l_i(x, t-1)$ is the unfulfilled demand in the previous period.[12] As in Silverberg (1987), the selective bench-mark is the industry average competitiveness:

$$E^1_m(t) = \sum_{i=1 \ldots n} E_i(t) \cdot f_i(t)$$

that is, the average of individual levels of competitiveness weighted by market shares (f_i).

In sector 2, individual levels of competitiveness are

$$E_j(t) = -\omega_4 \cdot ln(p_j(t)) - w_5 \cdot ln(l_j(t-1)) \quad (8.22)$$

and $E^2_m(t)$ is the analogous average competitiveness in the final good industry.[13]

Moreover, notice – not surprisingly – the similarity between equation 8.21 (the competitiveness of machines producing firms) and equation 8.17 earlier, expressing the evaluation functions of different machines/machine suppliers by would-be customers. In fact, the arguments of both the functions are 'objective' supply-side parameters (labour productivity allowed by the machine, prices and degrees of 'rationing') which differentiate producers. However, the actual competitiveness of each producer is also the outcome of varying distributions of *heterogeneous users' skills and users' technological expectations*. These distributions, weighted by the demand shares of each customer, determine the $\omega(t)$s in equation 8.21.[14] Each market share, f_i, changes according to a non-linear process

$$f_i(t) - f_i(t-1) = F^1(E_i(t-1) - E^1_m(t-1)) \quad (8.23)$$

Indeed, in the structure of the simulation exercises presented here, the F function itself is constructed to be only the *ex post* aggregate outcome of (mistake-ridden) evaluation and choice rules by individual buyers.

Similarly, market shares for the final good change as a function of the difference between individual and average competitiveness

$$f_j(t) - f_j(t-1) = F^2(E_j(t-1) - E_m^2(t-1)) \tag{8.24}$$

In analogy with Silverberg, Dosi and Orsenigo (1988) we assumed the following adjustment dynamics of individual prices. In sector 1

$$\begin{aligned} ln(p_i(x,t)) - ln(p_i(x,t-1)) = \\ \psi_1 \cdot (ln(p_i^d(t)) - ln(p_i(x,t-1)) + \\ \psi_2 \cdot (E_i(t-1) - E_m^1(t-1)) \end{aligned} \tag{8.25}$$

and similarly in sector 2

$$\begin{aligned} ln(p_j(t)) - ln(p_j\,(t-1)) = \\ \psi_3 \cdot (ln(p_j^d(t)) - ln(p_j(t-1)) + \\ \psi_4 \cdot (E_j(t) - E_m^2(t)) \end{aligned} \tag{8.26}$$

That is, changes in actual prices when expressed in logs (different among firms, *also in the homogeneous good industry*) are influenced by desired (mark-up) prices, but they are also track the relative competitiveness of each firm in the market. The ψs are system parameters whose relative values capture the nature of the *product-market regime*: for example, a highly competitive regime is defined by relatively high ψ_2 and ψ_4 and near zero ψ_1 and ψ_3. The opposite applies to quasi-monopolistic markets.

Aggregate demand variables are emergent properties resulting from the microeconomic decision rules and interactions described so far. However, *ex post*, they are simply the sum of the corresponding realized decisions of individual units. Aggregate gross and net investments in machines (respectively $IE^T(t) + IS^T(t)$ and $IE^T(t)$) are the sum in j of individual investments. The same applies to employment. With fixed coefficient technologies and constant returns to scale, employment in the final good sector ($L^2(t)$) sums over the actual outputs compounded by labour productivities of each agent. In sector 1, employment is given by the sum of production employment and research employment (recalling that $I_i(t)$s – investments in research – are measured in labour: $L^2(t)$).

Of course, total employment is simply the sum over the two sectors $L^T(t) = L^1(t) + L^2(t)$.

Aggregate ('national') income, $Y^T(t)$, follows the standard accounting identities (it equals total production which equals total expenditure plus 'involuntary' changes in inventories). Finally, on the labour market we assume that wages (expressed in logs) change according to

$$\begin{aligned} \Delta ln(w(t)) = \zeta_1 \cdot \Delta ln(p_m^2(t-1)) + \zeta_2 \cdot \Delta ln(\Pi^T(t-1)) \\ + \zeta_3 \cdot \Delta ln(L^T(t-1)) \end{aligned} \tag{8.27}$$

where p_m^2 is a price index for the final good and Π^T is the average labour productivity in the economy.

BEHAVIOURAL NORMS, TECHNOLOGICAL LEARNING AND REGIMES OF GROWTH: SOME PRELIMINARY RESULTS

Clearly, the structure of the model with its complex thread of positive feedbacks prevents analytical solutions so that inferences on its properties can only be drawn from simulation exercises. In a first illustrative simulation we assume 10 agents in each sector that start with identical technological competences, but different parameters in their search rules (search investments, and propensity to innovate and imitate in sector 1) and in their technological expectations ('confidence' in new products machines for the agents of both sectors, and learning expectations in sector 2: see Appendix I for a description of some parameters of the simulation). In the competitive process, market interactions determine variation in the relative size and profitability of each firm (they may also 'die' and, in sector 1 only, 're-enter' with some probability, which is actually equivalent to the birth of new firms).

Figure 8.4 presents the dynamics of notional opportunities (those generated at aggregate level by income changes that yield new discoveries in 'basic science') and of opportunities that actually become technological innovations. As the simulation shows, the innovative exploration of individual firms 'floats' in an expanding set of potential new 'technological paradigms'. However, their path-dependent processes of technological

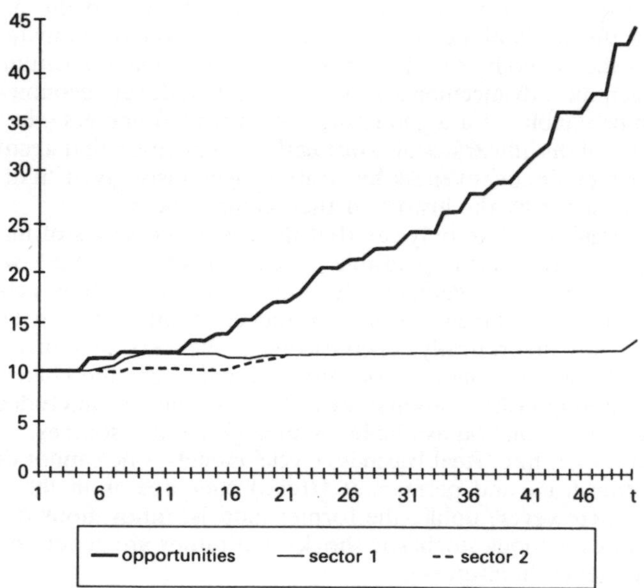

Figure 8.4 Distance between notional and realized opportunities

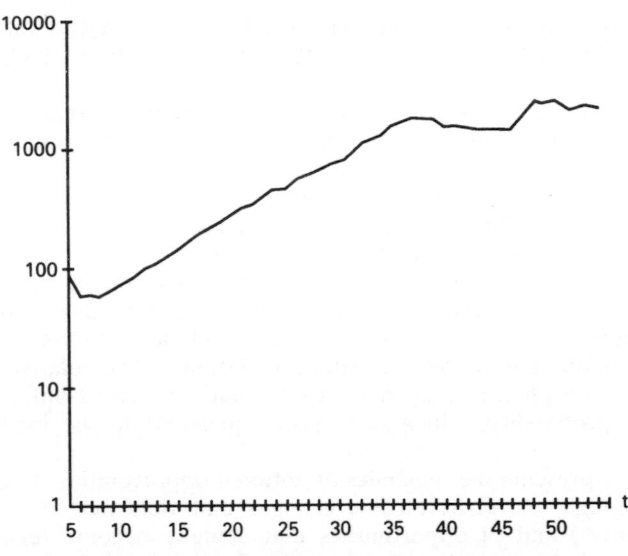

Figure 8.5 Income time series (logscale)

learning make them discover and develop only three of them, corresponding to the vertical steps along the otherwise horizontal line. Interestingly, this property appears to be quite resistant to changes in the parameters governing the probabilities of successful innovation by individual firms. Whenever the economy sustains a flow of new notional opportunities (that is, of 'scientific advancements') above a certain level, economic agents appear to be unable – for a good range of search parameters – to 'keep up' with them. All this illustrates also our earlier proposition that agents are not limited in their innovativeness by some exogenously given 'natural' constraint, but rather by the history of their competences.

At this stage, one can only say that the generated series of income and average productivity seem 'plausible' (see Figures 8.5 and 8.6): we conjecture that the model's aggregate dynamics might also show econometric properties similar to the empirical ones. Indeed, in our view, what should be surprising and, theoretically, encouraging is the very fact that relatively ordered patterns of economic coordination and aggregate change emerge without imposing either 'rational' *ex ante* consistency of microdecisions or exogenous constraints on available technologies and resources.

Note also that, like 'Real Business Cycle' models, one cannot distinguish between transitory and permanent (trend) components in the generated time series. However, unlike the former models, innovations do not take the form of exogenous stochastic shocks, but rather are generated endogenously by agents themselves.

Moreover, the transmission mechanism from innovation to aggregate dynamics is neither the intertemporal adjustment of consumption plans by a

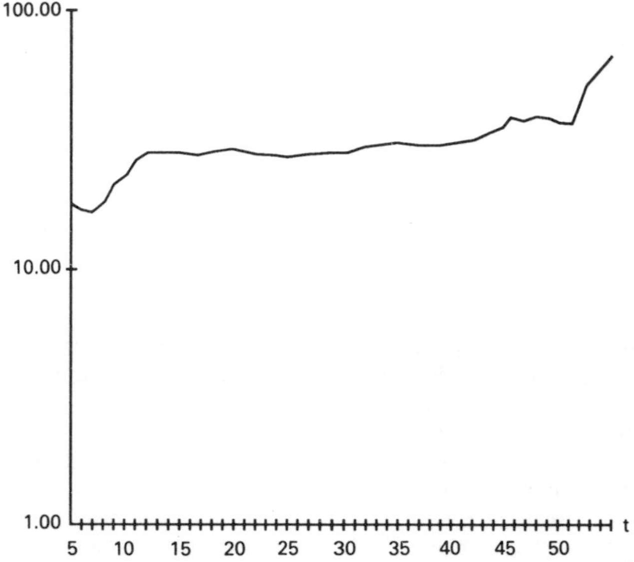

Figure 8.6 Labour productivity time series (logscale)

rational agent (see Long and Plosser, 1982), nor the optimal intertemporal distribution of investment plans (Schleifer, 1982). Indeed, in our model, innovation influences aggregate dynamics via two interrelated processes: *first*, via time-consuming diffusion among producers and users, and *second*, via the demand impulses that innovation and diffusion generate. The latter, 'Keynesian', feature of the model implies that, although (*endogenous*) 'shocks' have technological nature, the actual aggregate impact is determined by related *demand* shocks (via investment decisions, and levels of consumption).

Figure 8.6 shows the dynamics of labour productivity and illustrates also the possibility that the evolutionary dynamics may induce absolute decreases in average productivity even if best practice techniques may only improve (or remain the same): this is due to the learning lags of users and the possible fall in the market shares of early innovators. As with the results obtained by Silverberg, Dosi and Orsenigo (1988), innovators may sometimes be 'lambs' whose 'sacrifice' produces a learning externality for the whole system.

Finally, an 'impressionistic' consideration of both Figures 8.5 and 8.6 seems to hint at possibilities of major discontinuities in the relationship between income growth, productivity growth and, thus, also employment changes during different 'phases of growth' (compare, for example, the last 15 periods of both figures).

In a complementary work, Chiaromonte and Dosi (1990) analyse the essential role of *microdiversity* in aggregate growth. Contrary to the conventional microfoundations, one shows that the *form* of the distributions of microcharacteristics of the agents (technological competences and parameters of the decision rules) affect macrodynamics, *even for unchanged mean values*. For example, repetitions of the simulation shown in the above figures, assuming *identical* agents (with parameters equal to the mean values of the distributions as from the previous case), lead to very little technical progress and a long-term fall in aggregate income.

Collective learning and sustained income dynamics appear to find a necessary condition in technological and behavioural heterogeneity. Pushing the intuition to the extreme, what turn out to be 'mistakes' for individual agents might also represent positive externalities for the system as a whole.

Here, we want to focus upon the relationship between aggregate dynamics, on the one hand, and different institutional set-ups shaping behaviours and micro interactions, on the other.

In our model, these institutional features concern: (a) the *appropriability conditions of innovation*, as captured by the easiness of imitation in sector 1, and the speed at which private learning becomes a public externality in sector 2; (b) the *nature and parameters of decision rules* as they affect pricing, expansion, scrapping, R&D and so on; (c) the *nature of interaction processes* in the labour, product and financial markets.

As regards appropriability, corroborating Schumpeterian intuition, simulations not presented here show that some minimum appropriability threshold is a necessary condition for innovation. However, *variations* in appropriability, above a certain minimum threshold, have ambiguous effects on growth of income and productivity. For example, an increasing appropriability ensures the 'virtuous' feedback between innovation, profitability and growth for *individual* innovators, but may also slow down diffusion and, through that, *aggregate* growth in income and productivity.

More generally, in our model, 'institutions; are also the rules which govern adjustments in product and labour markets. Indeed, in the historical taxonomy of the various phases of capitalist development suggested by the French 'Regulation' approach, different 'regimes of growth' are also defined by different *average* adjustment parameters on the major markets. So, in an extreme archetype, a 'classical' ('competitive') regime is defined by the general absence of price-making power on the product markets and by the crucial dependence of real wage rates on unemployment level and changes. Conversely, the contemporary ('monopolistic') regime is defined – in that interpretation – also by various degrees of price-making behaviours on product markets and some (often union-led) indexation of wages on consumers' prices and on productivity increases. All that in our model implies that 'typical behaviours' in the product and labour markets would show up also in different adjustment parameters in equations 8.25, 8.26 and 8.27 above. In the highly constraint-free model defined here, the dependence of growth patterns upon particular institutional regimes would also appear through differences of dynamic patterns when variations occur in 'technological regimes' (that is, conditions of technological opportunity and appropriability) and in the average adjustment mechanisms on the various markets.

In fact, the simulation results shown earlier highlight a possible microfoundation to 'Kaldorian' positive feedback between (a) expanding technological opportunities; (b) endogenous demand growth (also via old-fashioned 'accelerator' mechanisms and growing wage-based consumption); and subject to (c) some institutionally specific constraints on income distribution (via some price making power on both labour and products prices).

What would happen if one changes average 'institutional' rules of adjustment, while leaving technological conditions unchanged? If one assumes 'competitive' price-making on either markets of products or labour (but not both), then, in both cases, a few simulation runs suggest that a rather flat trend of aggregate income appears. In our interpretation, the absence of *one of the institutional adjustment rules* deprives the system of one of the 'virtuous' positive feedbacks, and thus 'locks it' into a less expansionary dynamics (see Figures 8.7–10).

In the case of 'competitive wages' (that is, wages reacting only to employment changes) and oligopolistic product pricing (as the standard simulation: see Figures 8.5 and 8.6) a deflationary tendency is likely to appear. It is somewhat similar to that sort of mature 'capitalism depression' early envisaged by 'Keynesian' and 'structuralist' scholars such as Hansen, Steindl and Sylos Labini. Indeed, in this scenario, growth conditions seem to depend crucially upon the 'animal spirits' of entrepreneurs: if their demand expectations are high and so are their beliefs in innovative opportunities, aggregate growth is high too. But their 'optimism' in expanding technological opportunities and demand must *increasingly* compensate for unfavourable conditions of aggregate demand generation (in an old language, changes in income distribution make the 'Keynesian multiplier' shrink).

Conversely, 'competitive rules' on product markets (that is, prices reacting only to competitive conditions with no 'desired' mark-up) with 'monopolistic rules' on the labour market (as in the standard simulation from Figures 8.5 and 8.6) constrain the capability of firms to expand productive capacity and also to explore new technological opportunities (compare Figures 8.9 and 8.10 with Figures 8.7 and 8.8, and all of them with Figures 8.5 and 8.6); in the present model, this is the case nearest to the classical/neoclassical hyphoteses on some binding system-level constraint to firms expansion and aggregate income growth, via 'rigidities' in income distribution.

In simulations not shown here, we tried various combinations of 'competitive' adjustment parameters on both labour and product markets: that is, some highly stylized representations of a 'classical' regime of growth. Under these conditions, it turns out that the crucial parameters influencing the rate of long-term income growth are the investment propensities, as represented by the δs in equations 8.13 and 8.14: that is, in the empirical equivalents of the theoretical metaphor, the *norms and beliefs* governing the *propensity to accumulate* of business firms.

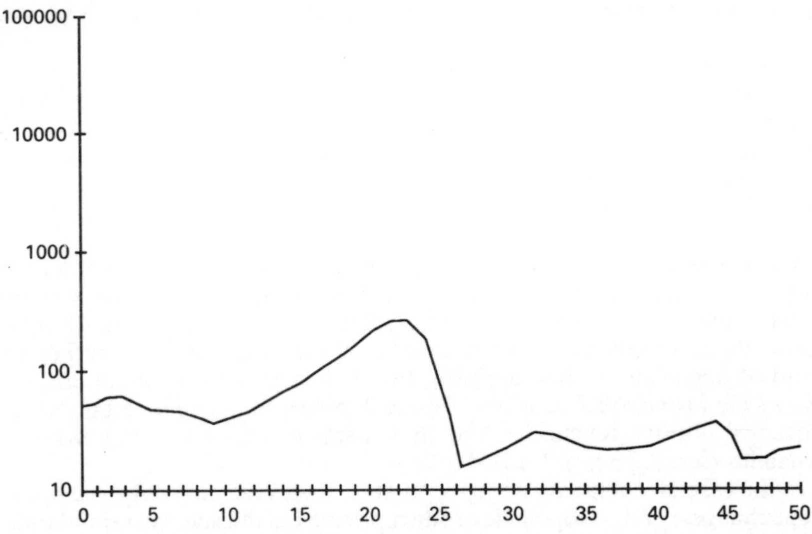

Figure 8.7 Income time series: 'classical employment' simulation (logscale)

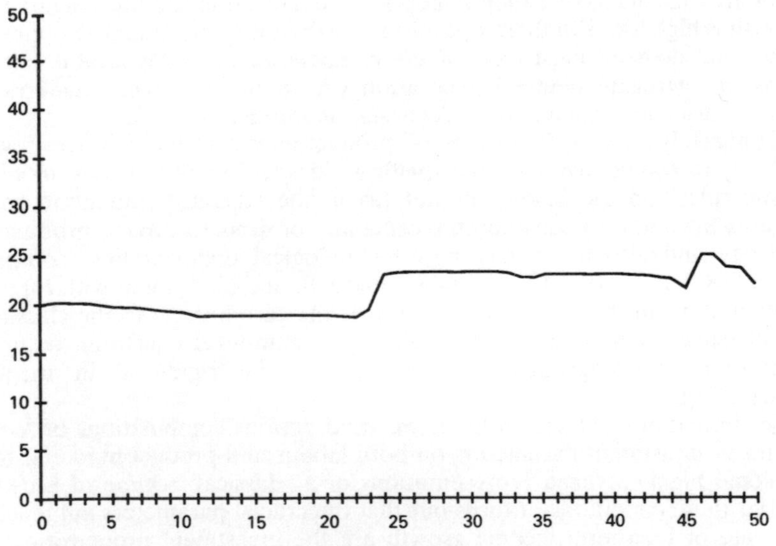

Figure 8.8 Labour productivity time series: 'classical employment' simulation

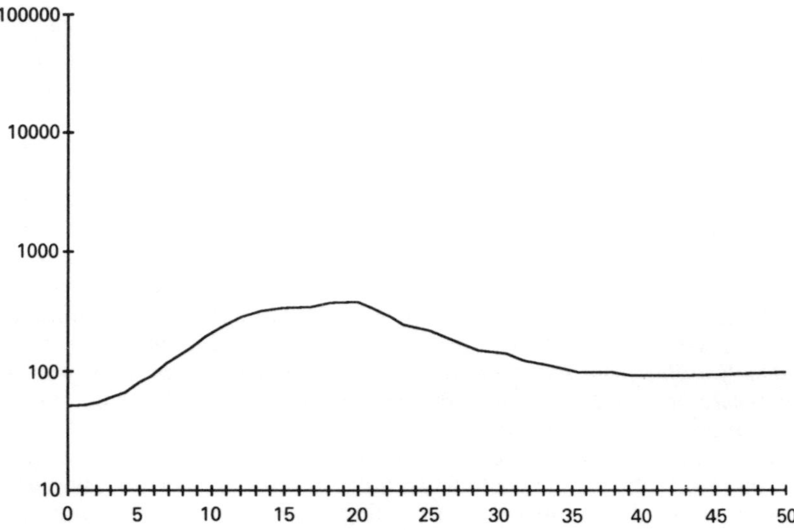

Figure 8.9 Income time series: 'competitive product-market' simulation (logscale)

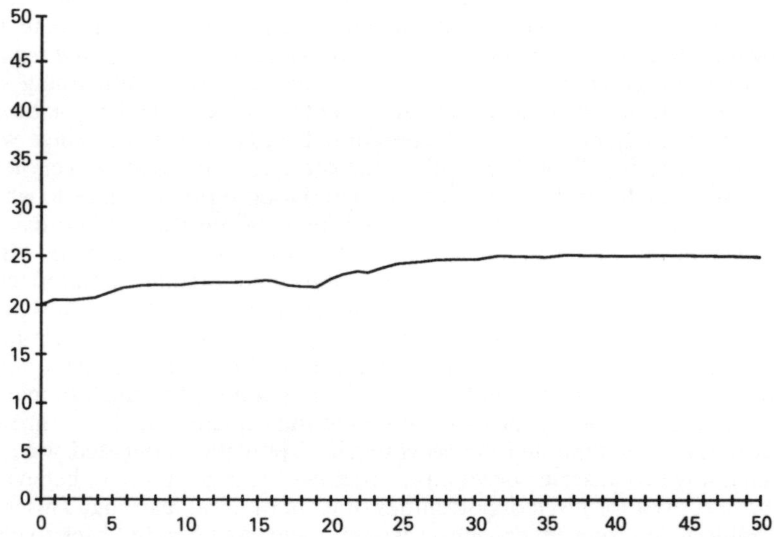

Figure 8.10 Labour productivity time series: 'competitive product-market' simulation

SOME CONCLUSIONS

The 'evolutionary' model presented in this work explores some of the properties of aggregate economic dynamics whereby, *first*, innovative opportunities are permanently and abundantly generated as direct outcomes of 'scientific' search and collective, indirect outcomes of individual interest-motivated behaviours; and, *second*, decision-guiding *norms* shape individual behaviour and collective adjustment mechanisms. Highly non-linear processes link individual strategies with aggregate outcomes. Moreover, in the *boundedly rational* environment depicted here, behavioural routines are rather inertial (at the very least, they change much more slowly than the notional technological opportunities which continuously emerge). Hence, the *rates and directions of individual learning might systematically differ from system-level learning*. And, also, one may well find *differences in the rates at which different institutional regimes exploit innovative and growth opportunities*.

We attempted to formalize a system dynamics as constraint-free as possible; therefore, the system could find a great (a priori unspecified) number of 'attractors' which it could lock-in according to its history and its technological and institutional characteristics. Our (admittedly very preliminary) investigation concerned some sort of mapping between technological *and institutional* set-ups, on the one hand, and system dynamics, on the other. In fact, a crucial feature of the model is to embed *endogenous* technological learning into specific behavioural and market institutions. We performed a few simulation exercises showing widely different aggregate outcomes of diverse conditions of microeconomic interaction which somewhat corrobate the conjecture that the various observed patterns of growth underlie *and require* highly specific combinations between collective technological learning and institutional forms of behavioural adjustments. As mentioned at the beginning, one would aim at a model wherein behavioural rules co-evolve with the system itself. Falling short of that, the present model is at the very least a 'theoretical tale' of the highly differentiated long-term outcomes to which varying combinations of technological learning and institutional norms can lead, whenever one abandons the modelling assumption of a world characterized by sophisticated calculating agents and binding 'natural' constraints to their actions. It can also be interpreted as a sort of *gedankenexperiment* (thought experiment) emphasizing some different system-level properties of learning and institutions (that is, different 'regimes') under increasing dynamic returns and much less than micro-optimizing rationality. In that, the model places the main burden of the explanation of 'why growth rates differ' over time and across countries upon the variegated ways by which innovative learning occurs. It also shows the importance of behavioural norms, even for *given* opportunities of technological learning. This is an old claim of development economics whose origin can be traced back to early political economists. The task of showing how different systems learn about institutions and rules has not been undertaken here: however, as difficult as it is, it seems within the reach of microfounded 'evolutionary' models.

Appendix I

Values attributed to the parameters defining the innovative microbehaviours in the simulation presented in the section on a self-organization model.

Parameters	*Means*	*Distributed between*:
Sector 1		
χ	0.15	0 and 0.3
μ	0.1	0 and 0.2
ζ	0.5	0.2 and 0.8
Sector 2		
a	0.2	0 and 0.4
b	5	3 and 7

The random distributions are of truncated normals around the above means.

Recall the meaning of the different parameters:

For sector 1 agents:

- χ is a firm-specific 'prudential' coefficient in the decision rule about radical innovations in manufactured machines.
- μ is the propensity to investments in technological search.
- ζ is the parameter that distributes research investments between innovation and imitation (propensity to imitate).

For sector 2 agents:

- a is the firm-specific learning expectation.
- b is the pay-back period parameter used in the scrapping rule.

Notes

1. In this and the following sections we shall present only some basic features of the model: more details are in Chiaromonte (1990).
2. Compare this view with that of an exogenous random arrival of 'technological shocks', as in Long and Plosser (1982), or the steady flow of innovations in Schleifer (1986). Nearer to our approach, *prima facie*, are Romer (1990) and Aghion and Howitt (1989), where the rates of innovation depend on the 'human capital' (or, more generally, labour) invested in exploration. However, in our model, the *abilities* in exploring are themselves an endogenous variable, and are, to some extent, firm-specific.
3. Cf. Freeman (1982), Winter (1987), Pavitt (1987) and Dosi (1988).
4. Cf. Aghion and Howitt (1989) who, however, use the expression to refer only to the substitution among different generations of innovation.
5. Of course, the 'Keynesianism' of Cambridge, England!
6. Although there is for individual borrowers (firms), as a function of their cash flow.
7. The restriction to the integer values of A_1 and A_2, while not influencing any of the conclusions, allows a positive probability that two or more different agents may produce/use the same machine (see below).

8. In the simulations of the penultimate section we have distinguished demand expectations on new typologies of machines, correcting them with a firm-specific parameter of 'confidence' in new technologies.
9. The following part of the model draws significantly on the earlier model in Silverberg, Dosi and Orsenigo (1988).
10. In its calculation, as in Silverberg, Dosi and Orsenigo (1988), one uses the actual skill levels: it would be too adventurous to throw away good equipment simply on learning expectations. Moreover, if the new machine belongs to a typology not yet adopted by j, the initial skill level is the public one $s_p\ (x, t)$.
11. In the model presented here there are no variable inputs other than labour, and overhead costs are independent from the levels of production.
12. Supposed equal to 0 if the currently manufactured machine is a new one.
13. Note, however, that, while the coefficients of the 'competitiveness' function of sector 1 agents are themselves functions of time, for sector 2 agents they are parameters (see below).
14. One cannot presume each producer would know the $\omega(t)$s. Rather, they result by aggregation over a heterogeneous population of customers.

References

Abramovitz, M. (1989) *Thinking about Growth* (Cambridge University Press).

Aghion, P. and Howitt, P. (1989) 'A Model of Growth through Creative Destruction' (Cambridge, Mass.: MIT Press), Dept of Economics, Discussion Paper.

Boyer, R. (1988a) 'Formalizing Growth Regime', in Dosi *et al.* (1988).

—— (1988b) *'Technical Change and the Theory of "Regulation"'*, in Dosi *et al.* (1988).

Chiaromonte, F. (1990) 'Processi di innovazione microeconomici e dinamiche macroeconomiche: un modello evolutivo bisettoriale' (Rome: Dept of Economics).

—— and Dosi, G. (1990) 'The Microfoundations of Competitiveness and their Macroeconomic Implications' (Paris: OECD/DSTI), presented at the Conference on 'Technology and Competitiveness', June.

Coricelli, F., Dosi, G. and Orsenigo, L. (1989) 'Microeconomic Dynamics and Macro-Regularities: An Evolutionary Approach to Technological and Institutional Change' (Paris: OECD/DSTI/SPR/89.7).

Dosi, G. (1984) *Technical Change and Industrial Transformation* (London: Macmillan).

—— (1988) 'Sources, Procedures and Microeconomic Effects of Innovation', *Journal of Economic Literature*, 26 (September).

—— and Egidi, M. (1990) 'Substantive and Procedural Uncertainty. An Exploration of Economic Behaviours in Changing Environments', *Journal of Evolutionary Economics*, 1.

—— Freeman, C., Nelson, R., Silverberg, G. and Soete, L. (eds) (1988) *Technical Change and Economic Theory* (London: Francis Pinter; New York: Columbia University Press).

—— and Orsenigo, L. (1988) 'Coordination and Transformation: An Overview of Structures, Behaviours and Change in Evolutionary Environments', in Dosi *et al.* (1988).

Freeman, C. (1982) *The Economics of Industrial Innovation*, 2nd edn (London: Francis Printer).

Greenwald, B. and Stiglitz, J. (1987) 'Keynesian, new-Keynesian and new-classical Economics', *Oxford Economic Papers*, 39.
—— and Stiglitz, J. (1988) 'Examining Alternative Macroeconomic Theories', *Brooking Papers on Economic Activity*, 11.
Grossman, G. and Helpman, H. (1991) 'Quality Ladders and Product Cycles', *The Quarterly Journal of Economics*, 106.
Kaldor, N. (1974) *Essays on Stability and Growth* (London: Duckworth).
Landes, D. (1969) *The Unbound Prometheus* (Cambridge University Press).
Long, J. and Plosser, C. (1982) 'Real business cycle', *Journal of Political Economy*, 91.
Lucas, R. (1988) 'On the Mechanics of Economic Development', *Journal of Monetary Economics*, 22.
Maddison, A. (1982) *Phases of Capitalist Development* (Oxford University Press).
Mansfield, E., Schwartz, M. and Wagner S. (1981) 'Imitation Costs and Patents: An Empirical Study', *Economic Journal*, 91.
Mokyr, J. (1990) *The Lever of Riches* (Oxford University Press).
Nelson, C. and Plosser, C. (1982) 'Trends and Random Walks in Macroeconomic time series', *Journal of Monetary Economics*, 10.
Nelson, R. and Winter, S. (1982) *An Evolutionary Theory of Economic Change* (Cambridge, Mass.: The Belknap Press of Harvard University Press).
Pavitt, K. (1984) 'Sectoral Patterns of Innovation: Towards a Taxonomy and a Theory', *Research Policy*, 13.
—— (1987) 'The Objective of Technology Policy', *Science and Public Policy*.
Romer, P. (1986) 'Increasing Returns and Long-Run Growth', *Journal of Political Economy*, 94.
—— (1990) 'Endogenous Technical Change', *Journal of Political Economy*, 98.
Rosenberg, N. (1976) *Perspectives on Technology* (Cambridge University Press).
—— (1982) *Inside the Black Box* (Cambridge University Press).
Schleifer, A. (1986) 'Implementation Cycles', *Journal of Political Economy*, 94.
Silverberg, G. (1987) 'Technical Progress, Capital Accumulation and Effective Demand: A Self-Organization Model', in D. Batten *et al.*, *Economic Evolution and Structural Change* (Berlin/New York, Springer Verlag).
——, Dosi, G. and Orsenigo, L. (1988) 'Innovation, Diversity and Diffusion: A Self-Organization Model', *Economic Journal*, 98.
Simon, H. (1984) 'On the Behavioural Foundations of Economic Dynamics', *Journal of Economic Behaviour and Organization*, 59.
Solow, R. (1984) *Growth Theory* (Oxford University Press).
Verspagen, B. (1990) '"New" neo-classical growth models and their relations to evolutionary theory of economic growth: an interpretative survey of some recent literature', Maastricht, University of Limburg, MERIT, Research Memorandum.
Winter, S. (1987) 'Natural Selection and Evolution', *The New Palgrave: A Dictionary of Economics* (London: Macmillan).

9 The Dynamics of Innovation and Diffusion with Competing Techniques

Willi Semmler

INTRODUCTION

Conventional economic theory usually assumes that information about new technologies and products is a kind of public good that can be costlessly and timelessly acquired. Accordingly, best-practice techniques are considered to be instantaneously adopted and implemented by existing firms. Much recent literature has departed from this view and regards the process of innovation and diffusion as more complex. Technological change, it is argued, takes place under conditions of competing technologies and imperfect (or only locally available) information, with innovation or diffusion costs for the new technology, returns to scale in adoption and learning, and high risk and uncertainty for the innovating firms. These conditions admit a wide range of outcomes in which the best and most efficient techniques do not always corner the market. These new approaches give a fresh and more realistic outlook to the study of innovation and diffusion processes. This chapter will identify three strands of this literature and will then utilize them to build a dynamic model of innovation and diffusion.

The first type of literature is known as the evolutionary approach to technical change. It revitalizes the classical and Schumpeterian theory of technical change[1] by borrowing from recent advances in mathematical biology to stylize the process of innovation and diffusion. An essential feature of this tradition lies in its behavioural assumptions. It expresses skepticism concerning the postulate of fully-fledged optimizing behaviour of economic agents. It argues instead that firms' behaviour in a changing environment is guided by limited information and bounded rationality. The informational requirements and computational complexity involved in fully rational behaviour would give rise to an unlimited cost of optimizing since, rigorously speaking, the assumption of rational behaviour implies that the cost of such optimizing behaviour should be part of the model's solution. Because it seems unreasonable to expect a full account of the cost of optimizing in a complex and uncertain environment, some rule other than optimization, such as satisficing or stopping rules, should be considered the foundation of firms' behaviour (see Conlisk, 1989). Limited information, risk and computational complexity would consequently give rise to uncertainty concerning which of the competing techniques will be implemented.[2]

A second line of research, related to the above approach, is present in the work done by Arthur (1987, 1988, 1989), who has recently rekindled the debate on innovation and diffusion by combining a probabilistic approach with the ideas of returns to scale and positive feedbacks (see also Arthur, Ermoliev and Kaniowvski, 1986).[3] In Arthur's view, innovation and diffusion of technology are rather accidental events which, from early in a technology's development, take place on the basis of competing methods or technologies, with the process of innovation having the properties of non-predictability. In this conception the competing techniques (or products) available 'may "compete" unconsciously and passively, like species compete biologically, if the adoption of one technology displaces or precludes the adoption of its rival. Or they may compete consciously and strategically, if they are products that are priced and manipulated' (Arthur, Ermoliev and Kaniowvski, 1986, p. 590). The reason why some innovations are rapidly implemented is explained on the basis of this probabilistic approach by reference to self-enforcing mechanisms. Technologies that gain an early lead by chance or historical luck may eventually dominate the market. Due to a positive feedback of stocks and flows in the innovation process,[4] an early implementation of a new technology can usually enjoy increasing returns to adoption resulting from learning by using, network externalities, scale economies in production, and increasing returns to information and skills. Insignificant early events may thus give rise to an initial advantage in adoption, and with adoption the advantage will be increased further. The new technology then may finally 'corner the market'.

A third strand of relevant research was initiated by Arrow (1962) and has subsequently been extended by Dasgupta and Stiglitz (1980a, 1980b) and Loury (1979), among others. In this view it is stressed that firms behave actively or strategically when introducing new technologies. In particular, firms are postulated to optimize when devoting resources to innovation. The effort made by firms to innovate (usually perceived as an R&D cost) is contrasted with revenue from the innovation: the revenue determined by a downward sloping demand curve. Special emphasis is given to uncertainty and risk involved in the dynamics of innovation and diffusion. Technological and market uncertainties arise in the relationship of research effort and the gains that can be captured from introducing the innovation. This in particular is due to the problem of timing, the interdependence of firms' decisions and possible rivals' reactions when new technologies are implemented.

This chapter incorporates features of all three of the above strands of inquiry. The analysis will, however, be restricted to studying models of deterministic dynamics.[5] A model of innovation and diffusion will be constructed whose dynamics is reminiscent of models in mathematical biology. More specifically, models of interacting populations will be referred to which develop three essential types of interaction: cooperative, predator–prey, and competitive (see Hofbauer and Sigmund, 1984). Two variations of our basic approach will be explored. In one variation, firms behave purely 'passively' when adopting a new technology; in the other, the innovating firms act strategically and optimize, although rationality is somewhat 'bounded' as it is more characteristic of the first approach.

In the remainder of the chapter we will survey some intra- and interindustry models of innovation and diffusion, formally close to the tradition of mathematical biology. These are non-linear models which work with either one, two or *n* variables. Next, we will explore a dynamic model of innovation and diffusion, with non-optimizing behaviour of firms. Here, cyclical diffusion processes and even limit cycles in diffusion processes are demonstrated to exist. Another section presents a variant of the model where optimizing behaviour is allowed for, but with a time horizon of innovating firms approaching zero. Optimizing takes place with respect to a return or profit function of the firm where the cost of innovation (or adoption of techniques) is contrasted with the returns.

As will be shown, with two state variables of our dynamic system, optimizing behaviour of the type characterized above can give rise to multiple equilibria, path-dependence in the sense of David (1990) and Arthur (1989), complex diffusion processes, and coexistence of diversities of techniques. Since there are several types of interactions, self-enforcing mechanisms may also eventually be turned into negative feedbacks which, in turn, may keep the market share of innovating firms bounded. The final section draws some conclusions on the subject.

SOME MODELS IN THE EVOLUTIONARY TRADITION

Since the different versions of our model build on approaches in mathematical biology and the perhaps not so well-known models arising from this tradition, we will begin by surveying four prototypes recently discussed in the innovation literature.

The simplest model of this tradition formalizing the process of innovation and diffusion is the logistic model which is theoretically well-developed and empirically often tested in case studies (see Mansfield, 1961; Davies, 1979; Metcalfe, 1981). It has been applied as an innovation and diffusion model for firms, industries and countries (Davies, 1979). It is a non-linear dynamical model in one variable, originally developed in biology for the limited growth of species (due to crowding effects) and in the analysis of the spread of epidemic diseases (Hirsch, 1984).

The dynamic equation for the diffusion process of an innovation reads as follows:

$$\dot{x} = \beta x(n - x) \tag{9.1}$$

with x the number of firms who have adopted the technology, n the equilibrium number of potential adopters of the new technology, β a reaction coefficient specifying the speed of diffusion of the new technology and the product xx the crowding effect (in models of interacting populations used to

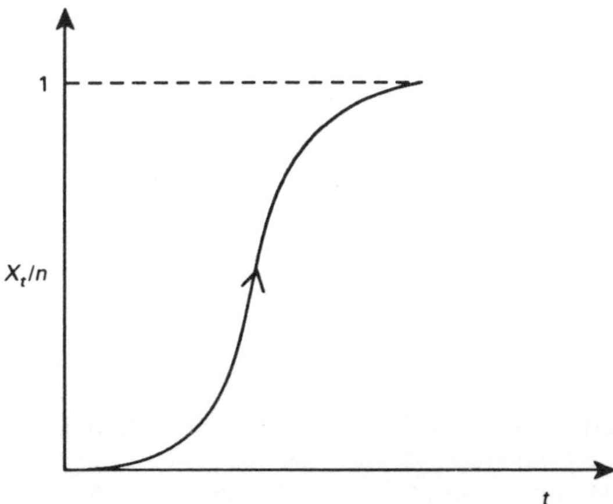

Figure 9.1 Logistic curve

indicate the competition of species for limited resources). The speed of diffusion β is usually, in the economic literature, seen to be dependent on (a) the growth of the industry, (b) competitive pressure, or (c) profitability differences (Metcalfe, 1981). The model implies that the number of potential adopters is exogenously given and not further specified endogenously.

The solution of the dynamic equation (9.1) gives the famous S-shaped curve as depicted in Figure 9.1 (for details of the analytical solution of (9.1), see Hirsch and Smale, 1974).

Other variations of the adoption process are possible. One might, for example, use time-discrete versions for the innovation process or include time delays in the differential equation such as

$$\dot{x} = \beta x(n - x_{t-1}) \tag{9.1'}$$

For an analytical treatment of these types of models, see Lauwerier (1986).

Moreover, interacting-species models often contain another species, y, which may, given limited resources, enforce the innerspecific crowding effect of the species x so that another term, for example, γxy, could be added to (9.1) or (9.1′). The dynamic properties are well known (see Luenberger, 1979; Lauwerier, 1986).

Second, Silverberg (1982) proposes a model with two variables, in which a third variable enters representing a new technology (or, in biological terms, a new species with higher selective potentials). Starting from a predator–prey system as proposed by Goodwin for the employment–wage share dynamics

$$\hat{v} = A - Bw$$

$$\hat{w} = -m + nv \tag{9.2}$$

with A, B, m and n constants, w the real wage and v the ratio employed to total labour force, and $\hat{v}$, $\hat{w}$ their growth rates (w/a = productivity, $A = 1/c - \beta$, $B = 1/ac$, and β a coefficient). An extension of the model so as to allow for the selection of a new technology and extinction of an old one gives rise to the following two-sector model:

$$\hat{v}_1 = A_1 - B_1 w$$

$$\hat{v}_2 = A_2 - B_2 w$$

$$\hat{w} = -m + (v_1 + v_2) \tag{9.2'}$$

with $v_1 = l_1/n$ and $v_2 = l_2/n$ and l_1, l_2, the respective employment in the two sectors and n the total employment. The old process has the subscript 1 and the new process the subscript 2, both temporarily coexisting in the economy. In biological models this corresponds to the appearance of a mutant species with higher selective potentials (see Hofbauer and Sigmund, 1984, Ch. 17). If $A_2/B_2 > A_1/B_1$ then the old process will be eliminated, v_1 approaches zero, and only the new process survives, giving rise to higher productivity and higher real wages. Note that there are, however, no further interaction effects between the old and new techniques (no cooperative or learning effects, or crowding effect).

Third, Silverberg, Dosi and Orsenigo (1988) constructed a model with further specified intra-industry interactions of firms and with technological varieties and behavioural diversities, which are postulated to exist persistently. The n firms within an industry are equipped with a disparity of relative competitiveness. Competitiveness is made dependent on the price that each firm can offer and the delivery delay of its product. As in Iwai (1984), the market share of a firm i can grow if its competitiveness is superior to the weighted average competitiveness of all other firms in the industry. In models of interacting populations, the higher growth rate of a species is due to its higher selective potentials. Similarly, in industry models firms will exhibit a higher growth rate of their market share with superior competitiveness. Here, crowding-out of firms – except one – would take place, since there are no other interaction effects among firms (no predator–prey relation or cooperative effect) which may prevent the firms from being eliminated.

In the model, firm i's market share grows at a rate $\hat{f}_i$. Thus the model reads as follows:

$$\hat{f}_i = A_0 \, (E_i - <E>) \tag{9.3}$$

with $E_i = -\ln p - A_{10} dd_i$ the competitiveness of firm i, dd_i the delivery delay, A_0, A_{10} constants, and $<E> = \Sigma f_i E_i$ the weighted average competitiveness. The dynamics of this model may, because of the missing interaction effects among firms, result in complete extinction of other firms. This has been shown for selection models of species (see Eigen and Schuster,

1979; Hofbauer and Sigmund 1984; see also Iwai, 1984, for a related dynamic model for firms' survival). In the model of Silverberg, Dosi and Orsenigo (1988), the extinction of firms is, however, counteracted by other dynamic feedback mechanisms (for details, see Silverberg, Dosi and Orsenigo, 1988). Due to those other feedbacks, market shares of firms are endogenously determined as a result of the model's characteristics (as hypothesized for models of Schumpeterian features; see Nelson and Winter, 1982). The competitive process converges towards differential market shares of the n firms in the industry.

Fourth, a more general model of technical change with process selection and extinction for m processes and n products is formalized in Flaschel and Semmler (1987, 1992). There, an innovation is represented by a selection of superior techniques. Inferior (existing) processes are eliminated through competition. The model also allows for product innovation and extinction. The basic structure of the model is an n-variable predator–prey dynamics. Although different in its economic content, it can be considered a generalization of the second model, as discussed above. Formally, this approach is elaborated in models of interacting species (see Rouche, Habets and Laloy, 1977).

More specifically, the Flaschel and Semmler model reads as follows. Let A and B represent (square) input–output matrices of a (constant returns) joint production system where workers' consumption goods are included in the input-matrix A. Its dynamics can then be formulated as follows:

$$\dot{x} = +\langle d^1\rangle\langle x\rangle(B - R^*A)\,'p' = +\langle d^1\rangle\langle x\rangle C'p' \tag{9.4}$$

$$\dot{p}' = -\langle d^2\rangle\langle p\rangle(B - R^*A)x = -\langle d^2\rangle\langle p\rangle Cx \tag{9.5}$$

with $\langle d^1\rangle$, $\langle d^2\rangle$ diagonal matrices of reaction coefficients, and x and p the output and price vectors, respectively. A dynamic version for output is stylized in (9.4) where excess profitability ($R^* = 1 + r^*$ being the average profitability) gives rise to output changes. In (9.5) excess demand, where $(1 + r^*)Ax$ denotes the vector of demand, gives rise to price changes.

If firms also take into account the sign of change of extra profits (or losses) when moving their capital between activities, then rising extra profits will speed up the growth rate of the respective sector, while falling extra profits will tend to reduce the growth effect of supernormal profits. We then get

$$\dot{s} = d/dt\ (B - R^*A)'p' = (B - R^*A)'p' \tag{9.6}$$

This vector shows the direction towards which extra profits (or losses) will change at a point in time t. Inserting (9.6) into (9.4), we obtain a new dynamic system

$$\dot{x} = +\langle d^1\rangle\langle x\rangle[C'p' + \gamma s] \tag{9.7}$$

$$\dot{p}' = -\langle d^2\rangle\langle p\rangle Cx \tag{9.8}$$

where $\gamma>0$ is an adjustment parameter. In matrix form we obtain

$$\hat{z} = <d>\begin{bmatrix} \gamma S & C' \\ -C & 0 \end{bmatrix} z \qquad \text{or } \hat{z} = <d>Q(\gamma)z \tag{9.9}$$

where $z = (x, p')'$, and $d = (d^1, d^2)'$. It has been shown in Flaschel and Semmler (1987, 1992) that system (9.9) is asymptotically stable.

Next, let us assume that we start with m techniques available for firms and n products, and postulate that our production model A, B (with $A^j \geqslant 0$, $B^j \geqslant 0$ for all processes $j = 1 \ldots m$) has an equilibrium $R^* > 1, x^*, p^* > 0$ which fulfils $p^* > 0$: that is, which is characterized by

$$p^*C \leqslant 0,\ Cx^* = 0,\ p^*Bx^* > 0 \tag{9.10}$$

Note that we allow in (9.10) for $x_j^* = 0$ where $pC^j < 0$ holds true (process extinction), but not $p_i = 0$ (no free goods). Then, given a system with $m \geqslant n$, the case of process extinction can be treated with an extended Liapunov function as discussed in Rouche, Habets and Laloy (1977, pp. 260–70). Asymptotic stability can be demonstrated to exist where inferior processes are eliminated and new processes implemented through the dynamics of competition as stylized in (9.7) and (9.8).[6]

On the other hand, product innovation and extinction can also be dealt with as a case where a modification of the 'law of excess demand' instead of the 'law of excess profitability' is to be introduced. Here, the dynamics can be written as

$$\hat{z} = <d>\begin{bmatrix} 0 & C' \\ -C & \delta U \end{bmatrix} z \tag{9.11}$$

with the matrix $U = -C<x><d^1>C'$. In the case of no inferior activities ($x^* > 0$), this dynamic may then be treated in the same way as we have treated the opposite case, $p^* > 0$. The adjustment will be quasi-globally stable and will now exhibit product extinction (see Flaschel and Semmler, 1987).[7]

Although the above generalized version of a model on technical change permits us to study the selection, extinction and stability of processes (and products) analytically, its shortcoming is that information about the new techniques (or the new set of products) is assumed to be obtained costlessly. There are no innovation or adoption costs involved in the diffusion of new processes (or products). With innovation or adoption costs the intra-industry dynamic among firms appears to be more complex and may give rise to the coexistence of a diversity of techniques. Two different versions of such a type of model will subsequently be explored.

Since most of the aforementioned studies are analytically solvable, they are quite helpful as reference models for a more complicated dynamic. Particularly, a more adequate modelling of the intra-industry interactions of firms with competing techniques, where firms face innovation or adoption costs, is desirable; the next two sections will deal with this.

INNOVATION AND DIFFUSION WITH NON-OPTIMIZING BEHAVIOUR

The subsequent model deals with intra-industry innovation and diffusion processes where firms exhibit non-optimizing behaviour. Following Conlisk (1989), existing firms may be divided into two classes: firms that produce with the existing technique and firms starting to produce with a new technique. The number of old firms switching over to the new technique is solely a function of the number of new firms already applying the technique. The increase in the number of firms operating the new technique will then have an unspecified crowding effect for all firms.

The model works with three types of interaction effects between the two types of firms: a basic predator-prey relation between the new and the old firms; a cooperative effect; and a competition (or crowding) effect. The predator–prey relation results from the fact that the new firms can grow only at the expense of the number of old firms. Predator–prey relations are one-sided in that innovating firms grow at the expense of old firms, but not vice versa. Old firms can grow faster than innovating firms only if the number of innovating firms is small. The crowding effect results from a price or mark-up squeeze due to the introduction of a new technique. An inverse demand function will be used here to specify this effect.[8] A kind of cooperative effect (spillover or learning effect) will help to keep the number of old firms away from zero, so that they are never completely extinct.

The dynamic result will be that old and new firms may coexist and their number may change cyclically.

The basic model, which resembles the sales-advertising model for two products (see Dodson and Muller, 1978; Feichtinger, 1989), reads as follows:

$$\dot{x}_1 = k - a(x_1, x_2)x_2 + \gamma\delta(x_2)x_2 - e\varepsilon(x_2)x_1 \tag{9.12}$$

$$\dot{x}_2 = x_2(a(x_1, x_2) - \delta(x_2)) \tag{9.13}$$

where x_1 and x_2 denote the pool of firms with old and new techniques respectively, and k is a constant, representing the number of firms starting with the old technique (which can be very small). The expression $a(x_1, x_2)x_2 = ax_1x_2x_2$ denotes the predator–prey interaction where the adoption of the new technique is supposed to take place proportionally to the product of x_1 and x_2x_2, a common assumption for the spread of information in sales—advertising models (see Feichtinger, 1989). This implies that as the number of firms applying the new technique grows, so does the accessibility of that technique for old firms. Note that the rate of decrease of old firms in (9.12) is equal to the rate of increase of new firms in (9.13).

In (9.13) the term $\delta(x_2) = \beta x_2 + \varepsilon(p(x_2))$ represents two effects: the natural rate of exit of firms x_2 which have failed when attempting to apply the new technique, and a rate of exit of firms of type 2 due to a crowding effect. The crowding effect results from an inverse relation between the price (mark-up)[9] and the number of firms of type 2. The crowding effect is specified as

$$\varepsilon(p(x_2)) = 1/p = 1/(\alpha/(x_2 + \phi)),$$

with $p = \alpha/(x_2 + \phi)$ the inverse demand function whereby ϕ keeps the rise of prices bounded. ϕ is chosen to be positive, since otherwise $p \rightarrow \infty$ if $x_2 \rightarrow 0$. Subsequently, we focus on the parameter ϕ as a critical bifurcation parameter in the dynamics.

On the other hand, the term $\gamma(\delta x_2)x_2$ may be regarded as an expression that helps to keep the number of firms of type 1 away from zero. It represents the number of firms – although probably small – that restart the business with the old technique and a spillover and learning effect for old firms. In addition, $e\varepsilon(x_2)$ stands for the crowding effect on old firms, since the price (or mark-up) resulting from the number of new firms also affects the survival of old firms.

Thus, the dynamic system (9.12) and (9.13) can be regarded as a natural extension of the logistic approach of (9.1)–(9.1′) because the additional dynamic equation, one for the firms with the old technique, will help to explain the potential number of adopters, n, as well as the crowding effect $(-\beta xx)$, which were unspecified in the dynamics (9.1)–(9.1′).

The computation of the Jacobian of (9.12), (9.13) gives more information about the local dynamic behaviour.[10] The elements of the Jacobian are:

$$J_{11} = -ax_2^*x_2 - e(x_2^* + \phi)/\alpha;$$

$$J_{12} = -2ax_1^*x_2^* + \gamma(2x_2^* + \phi)/\alpha - ex_1^*/\alpha;$$

$$J_{21} = ax_2^*x_2^*;$$

$$J_{22} = 2ax_1^*x_2^* - (\beta + (2x_2^* + \phi)/\alpha));$$

In general, it could be that $Tr\,J \geq 0$, but if x_2^* is much larger than x_1^*, then possibly $Tr\,J < 0$. The $Det\,J$ may switch sign, depending on ϕ. For small ϕ we may have $Det\,J < 0$, for increasing ϕ possibly $Det\,J > 0$, and for a still increasing ϕ $Det\,J < 0$ again. The equilibrium is turned from a saddle point to a locally asymptotically stable point and back to a saddle point.[11]

The simulation results, with varying parameter ϕ, representing the elasticity of demand (the impact of x_2 on the mark-up), revealed for small ϕ a saddle and for an increasing ϕ converging behaviour of the trajectories (for the employed parameters, see Appendix I). The converging cases are depicted in figures 9.2a and 9.2b. Whereas in Figure 9.2a we chose a = 0.1, ϕ

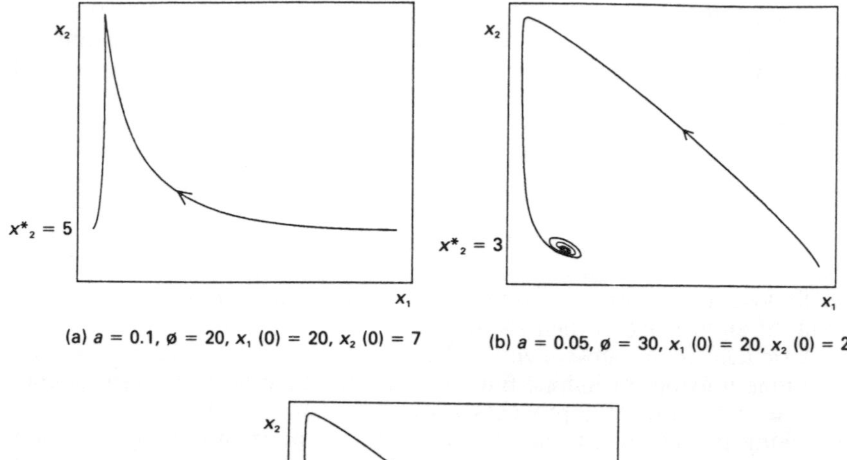

(a) $a = 0.1$, $\phi = 20$, $x_1(0) = 20$, $x_2(0) = 7$ (b) $a = 0.05$, $\phi = 30$, $x_1(0) = 20$, $x_2(0) = 2$

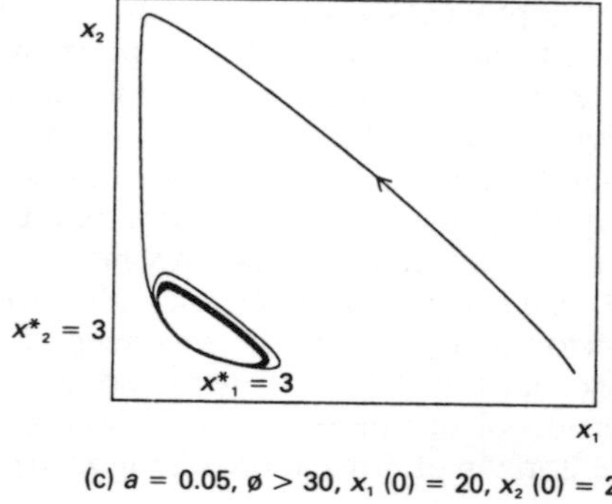

(c) $a = 0.05$, $\phi > 30$, $x_1(0) = 20$, $x_2(0) = 2$

Figure 9.2 Innovation and diffusion with non-optimizing behaviour

= 20, in Figure 9.2b the parameters are a = 0.05, ϕ = 30. Figure 9.2a depicts x_2 alone, in order to demonstrate the difference of our result to the logistic function (x_1 converges to 2). By further increasing ϕ slightly, however, the trajectories converge towards a limit cycle (see Figure 9.2c). A larger ϕ than used above turned the equilibrium again into a saddle (not depicted here).

INNOVATION AND DIFFUSION WITH OPTIMIZING BEHAVIOUR

We now want to allow for optimizing behaviour of firms in a model such as (9.12) and (9.13). Thus, firms behave actively. They invest in new technologies and expect a rate of return from that investment. We want to posit that the new technology will be either produced or acquired at a certain cost (an innovation or adoption cost).[12] The total cost for operating the new technology is assumed to be dependent on the effort spent to operate the new technology (independent of the number of firms) and a cost proportional to the number of firms operating it.

Furthermore, in the model (9.12) and (9.13), it did not appear to be realistic that the crowding effect of firms sets in (and continues to exist) when the share of firms with the new technique (x_2) is positive but very small. With a costly new technique, firms will most likely have an unprofitable period when the new technique is introduced, then face a period when they can capture a technological rent (transient surplus profit), and then finally lose their technological rent, with a subsequent crowding effect as a result of an increase in their numbers.[13]

Optimizing in the subsequent model does not take place with respect to a long time horizon, or infinite time horizon; the time horizon in the simple version of our model approaches zero. This can be viewed, as shown in Sieveking and Semmler (1990), as a limit case of intertemporal optimization when the discount rate approaches infinity.[14] Economically, the meaning of this is that firms heavily discount the future because of fierce competition, informational constraints, uncertainty and high risk about next period's revenue. For a theory of technical change, where one usually postulates a strong interdependence of firms' decisions, this limit case seems to be of considerable interest. A more extended version with intertemporal optimizing in an infinite horizon model where the revenues of future periods are discounted with a discount rate $\delta < \infty$ is discussed in Semmler (1991).

An optimizing version of our model, with a discount rate approaching infinity (and the time horizon of firms approaching zero) reads as follows:[15]

$$\operatorname*{Max}_{u} \{g(x_2, u); u \in \Omega\}$$

with $g(x_2, u) = \mu(x_2, u)x_2u - cu - c_0x_2,\ \mu = \alpha/(\phi + x_2u)$

s.t.

$$\dot{x}_1 = k - a(x_1, x_2)x_2 + \gamma\beta x_2 - e(\phi + x_2u)x_1 \tag{9.14}$$

$$\dot{x}_2 = x_2\,(a(x_1, x_2) + vg(x_2, u) - \beta) \tag{9.15}$$

where x_2 is again the number of innovating firms, and u a variable (control variable from a control space Ω) indicating the level of effort spent to produce or acquire the new technology.[16] The cost per unit of effort is denoted by c. The cost, cu, is independent of the number of firms,[17] and there is a cost proportional to the number of firms, c_0x_2. Thus, $cu + c_0x_2$ is the amount of resources that innovating firms have to devote to the innovation. The total cost of innovation can be conceived as the entry-cost to the new technology (for the innovating firms as a group). The term $\mu\,(\cdot)$ is here to be interpreted as the (net) price received for the product produced by the new technology, with the $\mu(\cdot)x_2u$ the (net) revenue.[18]

When firms maximize a technological rent, $g(\cdot)$, facing a revenue $\mu(\cdot)x_2u$ and the cost $cu + c_0x_2$, a positive rent will enhance the growth of their numbers. We thus assume here that excess profit, with entry proportional to excess profit, will increase the number of innovating firms. In (9.15), therefore, a term appears that determines the extent to which the technological rent can be transformed into the growth of type x_2 firms (see also Iwai, 1984). Thus a more complicated term than $\delta(x_2)$ in (9.13) will have to be added. This term is $vg(x_2, u)$ in (9.15), with v a constant.

The term $\gamma\beta x_2$ reflects the cooperative effect of x_2 on x_1 (learning by old firms;[19] and in addition, firms may commence again with the old technology after being eliminated by the dynamics (9.15)). The last term, $e(\phi + x_2u)x_1$, is the crowding effect for x_1 which is specified as $(e/\mu)x_1$, with e a parameter and μ the price determination from an inverse demand function. For simplicity, taking $\alpha = 1$ we have $\mu = 1/(\phi + x_2u)$. The optimal u^* can be derived by taking for $g(x_2, u)$, the derivative

$$\partial(x_2u/(\phi + x_2u) - cu - c_0x_2)/\partial u = 0$$

and solving for the optimal u^*, which depends on the parameters of the return function $g(\cdot)$ and x_2. In general, we get:

$$u^* = (\sqrt{(\phi x_2/c)} - \phi)/x_2 \tag{9.16}$$

For our numerical specification, the following properties for the return function hold:

$$g(x_2, u^*) < 0, \text{ for } x_2 < x_{2min};$$

$$g(x_2, u^*) > 0, \text{ for } x_{2min} < x_2 < x_{2max};$$

$$g(x_2, u^*) < 0, \text{ for } x_2 > x_{2max}$$

Numerical evaluations found that the return function $g(\cdot)$ is concave in x_2 within the above boundaries. The boundaries (x_{2min}, x_{2max}) are derived from the considerations shown in propositions 1, 2 and 3.

Proposition 1. There is a minimum number of firms x_{2min} in order to obtain a non-negative profit $(g(\cdot) = 0)$.

Note that $\mu(x_2, u) = 1/(\phi + x_2u)$ is falling in x_2 and u, and that u depends on x_2 (see (9.16)). Then the non-negative profit condition above means that there is an x_{2min} such that

$$(1/(\phi + x_{2min}u^*))x_{2min}u^* - cu^* - c_0x_{2min} = 0 \tag{9.17}$$

(The proposition is demonstrated using our numerical parameters: see Appendix II.1.) Note, however, that there is no other u that maximizes $g(\cdot)$, and that u cannot be set to zero even if $g(\cdot) < 0$. The latter part of the statement comes from the fact that there is a cost independent of u involved

(c_0x_2), which forces firms to operate the new techniques even if they have losses. With an optimal u losses are, however, minimized (see Appendix II.2 for a numerical example).

Proposition 2. There exist a number of firms $x_{2min} < x_2 < x_{2max}$ for which an optimal u^* generates a maximum of profit. Substitution of u^* from (9.16) into $g(x_2, u)$ results in $g(x_2, u^*) > 0$ for the interval $x_{2min} < x_2 < x_{2max}$ (see Appendix II.2 for the numerical computation).

Proposition 3. There is an $x_2 > x_{2max}$ for which the profit becomes negative. Thus there is a decline of the (net) revenue with increasing x_2 above a certain critical x_{2max} (see Appendix II.2 for the numerical calculation of $x_2 > x_{2max}$).

The equilibria of the system (9.14)–(9.15) with an optimal control u^* as defined in (9.16) were found interactively; partly by computation and partly by studying the vector field of the dynamics of (9.14)–(9.15) in certain regions of the plane.[20]

We found three equilibria of the optimally controlled system (9.14)–(9.15). They are all in the $\mathbb{R}^2_+$. The equilibria, of course, depend on the parameter of the system. We obtained for $a = 0.01$ the equilibria:[21]

$$(E1):x_1^* = 20,\ x_2^* = 0;$$

$$(E2):x_1^* = 31,\ x_2^* = 0.2;$$

$$(E3):x_1^* = 1.2,\ x_2^* = 17;$$

The first equilibrium ($E1$), can easily be computed by using u^* from (9.16), setting in (9.15) $x_2^* = 0$ and solving for x_1^*. The other two equilibria are found by studying the trajectories and the vector fields.

The computation of the Jacobian of the optimally controlled system (9.14)–(9.15) shows that

$$J_{11} = -ax_2^{*2} - e(\phi + x_2^*u^*);$$

$$J_{12} = -2ax_1^*x_2^* + \gamma\beta - ex_1(u^* + x_2\partial u^*/\partial x_2);$$

$$J_{21} = ax_2^{*2};$$

$$J_{22} = 2ax_1^*x_2^* - \beta + v\partial g(x_2, u)/\partial x_2.$$

For the chosen numerical values (see Appendix II.1) in numerical evaluations we obtained $\partial u^*/\partial x_2 < 0$ and $\partial vg(x_2, u)/\partial x_2 > 0$ for $x_2 \leq 7$ and $\partial vg(x_2$ $u)/\partial x_2 < 0$ for $x_2 > 7$.

In the local dynamics of the equilibria ($E1$)–($E3$), ϕ did not play an important role as a bifurcation parameter. The speed at which the new technology diffuses, represented by the parameter a, was an important parameter affecting the local dynamics at ($E1$)–($E3$), and so proved the level of x_2 to be important.

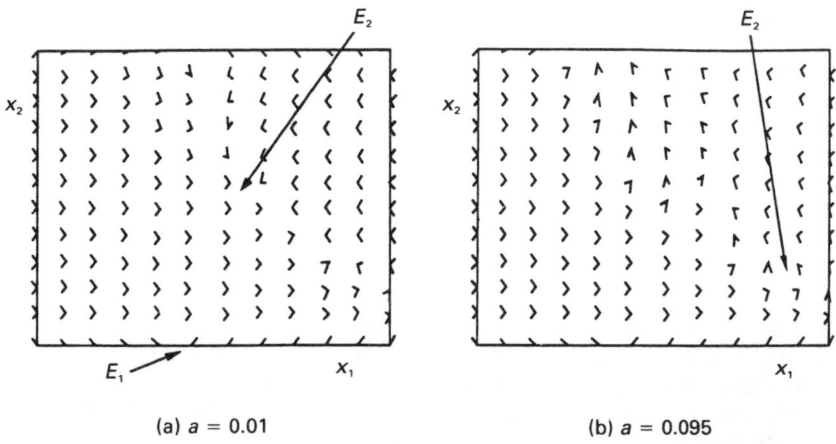

Figure 9.3 Vector field for the local dynamics about ($E1$) and ($E2$).

The local dynamics of ($E1$) can easily be computed since there $J_{12} = 0$. Since, due to $\partial vg(\cdot)/\partial x_2 > 0$, for small x_2 we possibly get $J_{22} > 0$ with $a > 0$, and therefore $TrJ > 0$ (or <0). Then we have Det < 0 (or >0). In the first case ($E1$) would be a saddle point; in the second case an attractor. Through the study of the local vector field, reported below in Figure 9.3, we have found that ($E1$) appeared to be mostly a saddle.

As can be seen in Figure 9.3a, although ($E1$) appears to remain a saddle with $a = 0.095$ the equilibrium ($E2$) alters to a repeller when the parameter a increases.[22] For an $a < 0.095$ it was found that trajectories starting close to ($E1$) or ($E2$) were always attracted by ($E2$). The low level equilibrium for the new technology was predominantly an attractor for initial conditions starting in the vicinity of it. When it began to be unstable will be discussed next.

In general, the stability properties of equilibrium ($E2$) also depend on the parameter a. With $J_{21} > 0$, due to $a > 0$, for large x_1^* the trace possibly becomes greater than zero and the Det $J > 0$. Thus ($E2$) might be an attractor, at least for certain values of the parameter a. When the parameter a was increased to $a = 0.095$, then ($E2$) became a repeller.

Figure 9.4a shows the trajectories for the case when $a = 0.01$ and initial conditions $x_1 = 33$, $x_2 = 0.3$. In this case the trajectories converge towards our equilibrium ($E2$):$x_1^* = 31$, $x_2^* = 0.2$. Due to the fact that $g(\cdot)$ switches sign when x_2 increases, ($E2$) is an attractor when the start value of x_2 is small. This demonstrates that the innovating firms do not get off the ground when the initial conditions (particularly for x_2) are close to the second equilibrium ($E2$). This apparently results from the innovation or adoption costs. Thus the new technology can get stock at a very low market share compared to the firms operating the old technology.

Figure 9.4b shows the trajectories with $a = 0.095$ and with initial conditions $x_1 = 33$, $x_2 = 0.3$. Now with a greater speed of diffusion of the new technology, the low-level equilibrium for x_2 ($E2$) becomes a repeller and ($E3$) has expanded its region of attraction. Thus, with higher adjustment speed for the new technology (higher a) the market share of the innovating firms may become dominant (since ($E3$) is now an attractor).

The third equilibrium ($E3$) is the most interesting one since it is comparable to the outcome of the logistic function discussed in the third section of this chapter. Note however that because of different functions and parameters, the equilibrium values must be different. The study of the equilibrium ($E3$) showed that it is an attractor for not too large values of the parameter a ($a < 0.095$)[23] and initial conditions $x_1 > 45$, $x_2 > 1$ (with x_2^* sizeable, we predominantly have $TrJ > 0$ and $\mathrm{Det}J > 0$ at ($E3$)).

As Figure 9.4c shows, with initial conditions $x_1 = 50$, $x_2 = 2$ and $a = 0.01$, the trajectories converge to ($E3$). The high-level equilibrium for the new technology dominates and the competing old technology loses its market share, although it can still maintain, contrary to the predictions from a logistic approach, a small market share. For a greater speed of diffusion, for $a = 0.085$, the trajectories become irregular (not depicted here) and finally, with $a = 0.095$, and the same initial conditions $x_1 = 50$, $x_2 = 2$, the trajectories divert to the low-level equilibrium ($E1$). This is demonstrated in Figure 9.4d. Other simulations, with a parameter value of a smaller than 0.05 and initial conditions closer to the high level equilibrium of x_2, showed that the equilibrium ($E3$) remains stable but the region of attraction became smaller.

These computer studies are quite illuminating. They demonstrate the possibly quite complex outcome of innovation and diffusion processes with innovation or adoption costs. Our studies indeed indicate that a slight change of the speed of diffusion, expressed by the parameter a (or of initial conditions), can give rise to a quite different behaviour of the trajectories (the trajectories displaying the market share of the different technologies). A change of the parameters or a change of initial conditions (starting number of firms operating the old and the new technologies) can give rise to low- or high-level equilibria of innovating firms. The change of the trajectories after altering the parameter a and initial conditions can stem either from a change of the stability properties of the (previously) attracting equilibrium or from the change of the isoclines of the equilibria. In both the former as well as the latter case the trajectories can end up in quite different equilibria.[24]

SOME CONCLUSIONS

This chapter has reviewed dynamic models for process innovation and diffusion which are akin to the evolutionary models originally developed in mathematical biology. In general, these models can encompass a dynamic

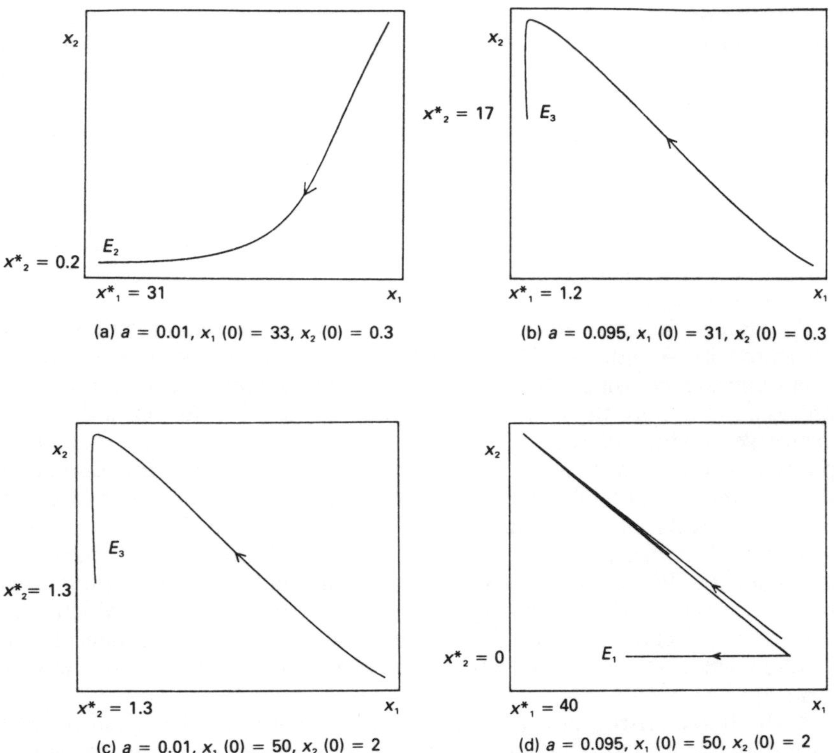

Figure 9.4 Trajectories of market shares of old and new technologies

interaction of three types: competitive, predator–prey and cooperative. These types of interaction have been shown to be useful for framing an approach to analyse technical change as process innovation and diffusion. To some extent the dynamics of product innovation could also be studied. In contrast to standard models of diffusion processes which utilize the logistic approach with a unique number of dominating firms (the innovators), our proposed model operates with two groups of competing firms: firms using the old technique, and innovating firms introducing the new technique. As demonstrated, more general interaction effects among firms (the predator–prey, learning or spillover, and crowding effects) appear to be desirable ingredients in the modelling of the dynamics of technical change. The proposed models of the previous two sections exhibit these features.

In addition to those types of interaction, in the last section optimizing behaviour of innovating firms and innovation or adoption costs of firms were explicitly introduced. When innovation or adoption costs, prices and transient technological rent are formally introduced, interesting new results arise. Because of the quite intricate interaction of the different types of firm, trajectories exhibit considerable diversity. Innovating firms may succeed, but firms with the old technique may still coexist with new firms. Innovating firms may also dissipate, if the innovation or adoption cost is too high (or the number of innovating firms is too small). Thus not only high- but also low-level equilibria for the innovating firms can become dominant attractors. The market shares for the old and new techniques may also exhibit cyclical features.

As stated, in some respects our above model is reminiscent of that presented by Arthur (1987, 1988, 1989). There, however, positive feedbacks due to increasing returns to adoption always dominate the innovation process. With more than one variable involved in the diffusion, a different distinctive feature of the innovation process can arise. In our models a positive feedback mechanism for the new technology may indeed dominate the dynamics for a while. Yet, at a later period, due to the interaction of competing techniques (resulting from crowding effects for the new and learning effects for the old firms), the innovating firms' market share can again decline. In addition, if low levels of equilibria for the innovating firms become attractors this is equivalent to shrinking market shares which may lead to the accumulation of debt by innovating firms. The financial risk arising from it, in turn, may diminish the probability of survival of the new firms.[25]

Such characteristics of a non-linear model with multiple equilibria and path-dependence in innovation and diffusion processes, possibly giving rise to a complicated dynamic, should be of interest for empirical innovation and diffusion studies.

Appendix I Numerical Specifications of the Non-Optimizing Model

For the model with non-optimizing behaviour of firms, we chose for Figure 9.2a the following parameters

$$\phi = 20; a = 0.1; \alpha = 100; \beta = 0.05; \gamma = 0.05; k = 1; e = 0.03$$

For Figure 9.2b the same parameters were chosen, but $a = 0.05$; $\phi = 30$. Figure 9.2c exhibits a limit cycle with $a = 0.05$ and $\phi = 33$. The bifurcation arises, for a given parameter $a = 0.05$, with respect to the parameter ϕ which has been increased in the simulations from 30 to 33.

Appendix II Numerical Specifications of the Optimizing Model

1. For the simulations of the model with optimizing behaviour of firms in the fourth section, the following parameters were chosen:

$\phi = 5; c = 0.1; c_0 = 0.04; k = 1; v = 0.5; \gamma = 0.8; e = 0.01; \alpha = 1;$
$\beta = 0.2;$

the parameter a varied from 0.01 to 0.095; and there were different initial values of x_1, x_2.

2. For the above parameters it could numerically be shown that Propositions 1–3 hold. The optimal u is:

$$u^* = (\sqrt{\phi x_2/c} - \phi)/x_2;$$

For g we obtain

$$g = u^*x_2/(\phi + x_2\, u^*) - cu^* - c_0x_2;$$

For the above parameters chosen we get for $g(\cdot) = 0$ a $x_{2min} = 0.73$ (our lowest number of innovating firms to allow for non-negative profits). On the other hand, x_2 should be smaller than x_{2max} which is for our numerical values roughly 17. Thus, any $x_{2min} < x_2 < x_{2max}$ will generate for an optimal u^* a $g(\cdot) > 0$. For $x_{2max} = 17$ we get again $g(\cdot) = 0$. Greater x_2 will generate losses even for the optimal control u^*. This numerically demonstrates Propositions 1–3. For the given parameters the critical value at which $\partial vg(\cdot)\partial/x_2$ starts declining is $x_2 = 7$.

Note, however, that firms of type x_2 will still optimize even if losses occur, since optimizing in the presence of fixed costs minimizes losses. For example, let the minimum number of firms x_{2min} be 0.73, at which $g = 0$. If firms choose an activity level $u = 2$ we get a $g = -0.003$, and for a $u = 1$ we have a $g = -0.001$. For $x_2 = x_{2min}$ the optimal u^* gives $g(\cdot) = 0$. For $x_2 < x_{2min}$ or $x_2 > x_{2max}$, the optimal u^* minimizes losses (when firms are locked in).

Notes

This version of the chapter was prepared while I enjoyed the hospitality of the Economics Department at Stanford University. I would like to thank Brian Arthur and Donald Harris for helpful discussions and suggestions.

1. For the evolutionary approach see Nelson and Winter (1982), Winter (1984), Prigogine (1976), Silverberg (1982), Silverberg, Dosi and Orsenigo (1988), and Bluemle (1989). Classical-oriented dynamical models can be found in Duménil and Levy (1987) and Flaschel and Semmler (1987, 1992). The evolutionary approach also refers to the classical writers (see Nelson and Winter, 1982, Chs 1–3).
2. Less uniformity and instead diversity with respect to behaviour and techniques would be predicted to exist empirically (see Silverberg, Dosi and Orsenigo, 1988; Winter, 1984; and for an interesting complete specification of such an approach, see Conlisk, 1989).
3. For a related view, see David (1990).
4. In a probabilistic version, the feedback between stocks and flows means that if there are n balls of m colours in an urn (stocks), the probability that a ball with a

certain type of colour will be added to the urn (flows) is proportional to the number of balls with the same colour already in the urn.

5. Risk and uncertainty will be referred to in connection to a discount rate to be discussed later.
6. Extensive proofs and computer simulations about how new techniques are implemented and old techniques eliminated are given in Flaschel and Semmler (1987, 1992).
7. For a single-product system – a simplified case of a joint production system – the dymanics with process and product innovation and extinction as proposed above and its stability results immediately apply if the output matrix B is replaced by the identity matrix I.
8. For a similar use of an inverse demand function in the context of an innovation model, see Dasgupta and Stiglitz (1980a, 1980b).
9. We assume here, without specifying further the cost of innovation, that the price and mark-up will be a declining function of the number of firms operating the new technique. More specific formulations are given in the next section.
10. Simulation techniques, studying the vector field in the plane, showed for our system that there might be several equilibria with different dynamic properties. In the context of the present, more descriptive, model we did not explicitly study the multiple equilibria and their stability properties in detail. It is more rewarding to undertake such an endeavour for the optimizing version of our model in the fourth section.
11. Note, if the dynamic system (9.12) and (9.13) has multiple equilibria, the more specific dynamics then depends on the properties of the isoclines for the different equilibria (see Sieveking and Semmler, 1990).
12. Thus financial resources have to be devoted to implement a new innovation. This can be perceived as R&D investment or adoption cost for the new technology. For related models, see Dasgupta and Stiglitz (1980a, 1980b), Conlisk (1989), and Foley and Lazonick (1990).
13. In our model it will be thus only for an initial stage that there are positive feedbacks, self-enforcing mechanisms or processes of cumulative causation working. Due to the different types of interaction, positive feedbacks may be turned into negative ones.
14. In Sieveking and Semmler (1990) it is shown that maximizing a value function for a discount rate δ approaching infinity, the trajectories x_δ approach the trajectories x of a system with time horizon approaching zero.
15. Optimization of the subsequent type will still be characterized by 'bounded rationality' in the sense that high risk next periods, limited information and computational complexity do not allow the agents (firms) to employ intertemporal optimization. One, however, does not have to stick to this limit case of an optimization problem. The time horizon can be extended. The relation of a model with time horizon shrinking to zero (static optimization problem) and a model with a finite discount rate (dynamic optimization problem) is elaborated in Semmler (1991).
16. The effort u could be interpreted as the amount of 'technological knowledge' (or information) obtained. The 'technological knowledge' could be thought of being created by (a) hiring engineers, (b) running research laboratories or (c) purchasing information on new technologies. In principle, however, the effort spent might not always lead to more technological knowledge and successful implementation of a new technology. There is considerable uncertainty and risk involved (see Arrow, 1962). Yet a stochastic version of the above model can

easily be formulated along the lines of Dasgupta and Stiglitz (1980a). We could write *huc* where the variable *h* represents a stochastic variable. It denotes the 'probability of success' for a unit of effort spent. We have kept $h = 1$.

17. We assume here that the new technology (new knowledge) is used jointly by firms x_2 so that the cost of acquiring the technology is independent of the number of firms of x_2 type. When the technology, however, is operated by the firms, the output resulting from it is posited to be proportional to the number of firms. Technological knowledge is a kind of public good for the cooperating firms; see also Romer (1990).
18. Note that we have chosen here a formulation of the (net) price (mark-up) where the price is proportional to cost. A more complicated cost function (with first decreasing then increasing cost) could easily be included in the return function: see Dasgupta and Stiglitz (1980a and b).
19. Learning by old firms might mean that they improve their production processes when information about the new technology spreads and the competitive pressure from the new technology increases. In our context it is reasonable to posit that information about the new technology leaks out more the larger the number of firms applying it.
20. The study of the dynamics of (9.14)–(9.15), the trajectories and the vector field, has been pursued by employing DS (Dynamical Systems). We do not exclude the possibility that there might be further equilibria. For a more detailed study of this problem, see Semmler (1991).
21. The parameters for our study of the equilibria and the dynamics are reported in Appendix II.1.
22. In the subsequent part we want to explore the dynamic effects of a change in the diffusion speed, depicted by the parameter *a*. Empirical studies for the diffusion speed of new products among countries is usually around 0.02 (see Jovanovic and Lach, 1990). Presumably the diffusion speed for new processes within industries is higher. We will work with a parameter size for the diffusion speed of less than 0.1.
23. We did not, however, explore extensively bifurcation values of other parameters involved, since we were mainly interested in the diffusion speed.
24. An analytical method to study problems of a dynamics with multiple equilibria, some of them with saddle point properties, is provided in Sieveking and Semmler (1990).
25. This financial aspect of innovation and diffusion processes appears to be an interesting topic for future research (see also Foley and Lazonick, 1990). Particularly in the context of models with returns to scale in adoption, the finance aspect is important. Since in the case of returns to scale firms are required to have a large market share to operate profitably, they may have to run temporary deficits and to borrow against the future before they dominate the market. This may turn out to be a quite risky financial operation. Such a problem is also discussed in Semmler (1991).

References

Arrow, K.J. (1962) 'Economic Welfare and the Allocation of Resources for Innovation', in R. Nelson (ed.), *The Rate and Direction of Inventive Activity: Economic and Social Factors* (Princeton, NJ: Princeton University Press).

Arthur, W.B. (1987) 'Competing Technologies: An Overview', working paper, Center for Economic Policy Research, Stanford University.

Arthur, W.B. (1988) 'Self-Reinforcing Mechanisms in Economics', in P. Anderson, K. Arrow and D. Pines (eds), *The Economy as an Evolving Complex System* (Reading, Mass.: Addison-Wesley).

—— (1989) 'Competing Technologies, Increasing Returns, and Lock-In by Historical Events', *Economic Journal*, 99, 394.

—— Ermoliev, Y.M., and Kaniowvski, Y.M. (1986) 'Path-Dependent Processes and the Emergence of Macro-Structure', *European Journal of Operations Research*, 30.

Bluemle, G. (1989) 'Wachstum und Konjunktur bei Differenzgewinnen-Ein Schumpeter-Model der wirtschaftlichen Entwicklung', in H.J. Ramser and H. Riese (eds), *Beitraege zur Angewandten Wirtschaftsforschung* (Berlin: Springer-Verlag).

Conlisk, J. (1989) 'Optimization, Adaptation, and Random Innovations in a Model of Economic Growth', University of California, San Diego, mimeo.

Dasgupta, P. and Stiglitz, J. (1980a) 'Industrial Structure and the Nature of Innovative Activities', *Economic Journal*, 90.

—— (1980b) 'Uncertainty, Industrial Structure and the Speed of Innovation', *The Bell Journal of Economics*, 11, 1.

David, P. (1990) 'The Dynamo and the Computer: An Historical Perspective on the Modern Productivity Paradox', *American Economic Review*, May.

Davies, S. (1979) *The Diffusion of Process Innovation* (Cambridge University Press).

Dodson, J. and Muller, E. (1978) 'Models of New Product Diffusion through Advertising and Word-of-Mouth', *Management Science*, 24.

Dumenil, G. and Levy, D. (1987) 'The Dynamics of Competition: A Restoration of the Classical Analysis', *Cambridge Journal of Economics*, 11, 2.

Eigen, M. and Schuster, P. (1979) *The Hypercycle: A Principle of Natural Selforganization* (Heidelberg/New York: Springer-Verlag).

Feichtinger, G. (1989) 'Hopf-Bifurcation in an Advertising Diffusion Model', University of Vienna, mimeo.

Flaschel, P. and Semmler, W. (1987) 'Classical and Neoclassical Competitive Adjustment Processes', *The Manchester School*, March.

—— (1992) 'Classical Competitive Dynamics and Technical Change', in E. Nell and D. Laibman (eds), *Beyond the Steady State* (London/New York: Macmillan).

Foley, D. and Lazonick, W. (1990) 'Corporate Takeovers and the Growth of Productivity', Working Paper Series, no. 91–01, Department of Economics, Barnard College.

Goodwin, R.M. and Punzo, L.F. (1987) *The Dynamics of a Capitalist Economy* (Boulder, Col.: Westview Press).

Hirsch, M.W. (1984) 'The Dynamical System Approach to Differential Equations', *Bulletin of the American Mathematical Society*, 11, 1.

—— and Smale, S. (1974) *Differential Equations, Dynamical Systems and Linear Algebra* (New York: Academic Press).

Hofbauer, J. and Sigmund, K. (1984) *Evolutionstheorie und Dynamische Systeme* (Hamburg: Paul Pavey Verlag).

Iwai, K. (1984) 'Schumpeterian Dynamics. An Evolutionary Model of Innovation and Imitation', Part I and Part II, *Journal of Economic Behavior and Organization*, 5.

Jovanovic, B. and Lach, S. (1990) 'The Diffusion of Technology and Inequality among Nations', C.V. Starr Center for Applied Research, no. 90–34, Department of Economics, New York University.

Lauwerier, H.A. (1986) 'Two Dimensional Iterative Maps', in A.V. Holden (ed.), *Chaos* (Princeton, NJ: Princeton University Press).

Loury, G. (1979) 'Market Structure and Innovation', *Quarterly Journal of Economics*, 93.

Luenberger, D.G. (1979) *Introduction to Dynamic Systems* (New York: Wiley).

Mansfield, E. (1961) 'Technical Change and the Rate of Innovation', *Econometrica*, 29.

Metcalfe, J.S. (1981) 'Impulse and Diffusion in Technical Change', *Futures*, 13, 5.

Nelson, R. and Winter, S. (1982) *An Evolutionary Theory of Economic Change* (Cambridge, Mass.: Belknap Press).

Prigogine, I. (1976) 'Order through Fluctuation: Self-Organization and Social Systems', in E. Jantsch and C.H. Waddington (eds), *Evolution and Consciousness* (Reading, Mass.: Addison-Wesley).

Romer, P. (1990) 'Endogenous Technical Change', *Journal of Political Econony*, 98, 2.

Rouche, N., Habets, P. and Laloy, M. (1977) *Stability Theory by Liapunov's Direct Method* (Heidelberg/New York: Springer).

Semmler, W. (1991) 'Optimal Inventive Investment and the Diffusion of Technologies', paper prepared for the Viennese Workshop on Dynamic Economic Models and Optimal Control, June.

Sieveking, M. and Semmler, W. (1990) 'Optimization without Planning: Economic Growth and Resource Exploitation when the Discount Rate tends to Infinity', University of Frankfurt/New School for Social Research, working paper.

Silverberg, G. (1982) 'Embodied Technical Progress in a Dynamic Model: A Self-Organizing Paradigm', in R.M. Goodwin, M. Krueger and A. Vercelli (eds), *Nonlinear Models of Fluctuating Growth* (Heidelberg/New York: Springer-Verlag), pp. 192–208.

——, Dosi, G. and Orsenigo, L. (1988) 'Innovation, Diversity and Diffusion: A Self-Organization Model', *Economic Journal*, 98.

Winter, S. (1984) 'Schumpeterian Competition in Alternative Technological Regimes', *Journal of Economic Behavior and Organization*, 5.

10 Learning and the Dynamics of International Competitive Advantage

William Lazonick

ENTERPRISES, NATIONS AND LEARNING

To produce goods and services requires knowledge. To acquire knowledge requires learning. Given a society's endowment of physical resources, the more it learns, the greater its productive potential. Learning can enable producers to use existing technologies more effectively. Learning can also enable producers to discover ways to substitute for those resources with which they are meagrely endowed. Learning, therefore, can both improve a society's productivity on the basis of available technologies and its ability to generate new technologies.

How does learning occur in a modern society, and what are the economics of the learning process? Why are some societies apparently better at learning than others, and how does superior learning contribute to superior economic performance? My purpose here is to draw on the history of modern capitalist development to provide a basic framework for answering these questions, with an emphasis on the role of social institutions in contributing to the learning process. Focusing on the business organization as a locus of learning, the ways in which the strategies and structures of business enterprises are central determinants of the rate and direction of the learning process will be outlined. This will be followed by a discussion of the regions or nations in which enterprises operate as loci of learning. Throughout this chapter, a theory of competitive advantage of enterprises and of nations as a historically relevant framework for analysing the economics of the learning process will be elaborated.

Learning is an economic problem because learning is rarely (if ever) costless. This proposition has, of course, been recognized by human capital theory, which was originally put forward to provide an explanation for how qualitative changes in human resources contribute to economic growth (see Schultz, 1968). But, besides the insight that people make costly investments in human resources on the expectation of returns, human capital theory lacks a relevant framework for analysing how and to what extent human resources are developed and utilized, and hence how and to what extent returns are generated. In elaborating human capital theory to distinguish between general and specific human capital, Gary Becker was concerned with the problem of allocating the burden of the costs of human capital investments between employees and employers in relation to the returns that the two parties to the employment relation could expect to reap from the investments (see Becker, 1968). But, wedded as he was to a theory that views market-coordinated outcomes as the ideal, Becker offered no perspective on how or to what extent long-term organizational ties between employers and employees can lead to the development and utilization of human resources in ways that outperform the perfectly competitive 'ideal' (for an elaboration of my views on these matters, see Lazonick, 1991a, Ch. 5).

If human capital theorists have recognized that learning is an economic problem, such has not been the case for all mainstream economists. Using the concept of 'disembodied technical change', some growth theorists have depicted learning as costless and its impact on productivity as automatic (see Solow, 1962). As used by neoclassical growth theorists, the concept of 'learning-by-doing' depicts producers, and hence society, as simply acquiring productivity-enhancing knowledge as by-products of productive activities. Such learning is viewed as costless in the sense that it would have been economically rational to undertake these productive activities even if learning were not to occur (see Arrow, 1962).

This conception of learning-by-doing implicitly assumes that the producer already has the knowledge of how to perform the relevant productive activity, and that productivity-enhancing improvement occurs through the repeated application of this knowledge. Yet even unskilled tasks require some instruction in 'best practice' techniques, if only to ensure that workers do not acquire 'bad habits': that is, to ensure that they do not learn to do tasks in ways that impede rather than enhance productivity (for an elaboration, see Lazonick and Brush, 1985). Indeed, the inexperience of the untutored worker can result in not only low levels of labour productivity, but also costly damage to materials and machinery as well as disruptions to the flow of work (Lazonick, 1990a). To attain the requisite learning, the business enterprise may invest in the training of workers, even if only by employing experienced workers to show the new recruits how to do their jobs.

Even for many basic types of work, therefore, learning is the result of the investments that the business enterprise decides to make. Indeed, for more sophisticated types of work that require intricate knowledge of the company's products and processes, the business enterprise may have to make substantial investments in human resources to ensure that key employees learn the requisite skills. Because investments in human resources are integrally related to an enterprise's investments in physical resources, an understanding of how, why and to what extent a business enterprise generates learning must begin with the investment strategy of the enterprise.

However, the analysis of investment strategy is not sufficient to comprehend the ability of the enterprise to generate learning. The extent to which an investment strategy generates learning depends on the organizational structure of the enterprise that makes the investments in human resources. Within a business enterprise, learning is a collective process of cognitive development in which the skills of the various individuals who participate in the specialized division of labour must be combined to achieve desired results. An investment strategy may be able to develop the skills of participants in the specialized division of labour, but it is the organizational structure of the enterprise that plans and coordinates the requisite combination of skills. As has been argued elsewhere, the greater the technological complexity inherent in an investment strategy, the more must the skill acquisition process be planned and coordinated (Lazonick, 1991, Chs 2–3).

The need for organizational coordination renders collective cognitive development a social as well as technical problem. In making investment decisions, strategic decision-makers within the enterprise may decide not to invest in the skills of certain types of worker if they think that the skills acquired through learning will be used to subvert rather than promote the goals of the enterprise. From the late nineteenth century until recently, most major American corporate employers were unwilling to invest in the skills of shop-floor workers because of the fear that the possession of skills would enable these workers to obstruct the quest for high-throughput production and low unit costs. Instead, these employers developed the skills of their managerial personnel, who in turn used their knowledge to search for machine technologies that would dispense with the need for shop-floor skills (Lazonick, 1990a, Ch. 6).

In other capitalist nations such as Britain and Japan, employers generally adopted different strategies concerning investments in human resources. During most of the twentieth century, British employers continued to rely on shop-floor workers to train other workers, as had been the custom in the nineteenth century. For the British, leaving skills on the shop floor was an alternative to the American strategy which focused on investments in managerial skills and the search for new machine technologies that could take skills off the shop floor. In contrast, Japanese employers neither left the acquisition of skills to shop-floor workers (as in Britain) nor sought to take skills off the shop floor (as in the USA). Rather, Japanese enterprises pursued a strategy of developing skills on the shop floor as complements to investments in human resource development within the managerial structure (Lazonick, 1990a; see also Gospel, 1991).

Why speak of 'British', 'American' and 'Japanese' employers, workers and enterprises? To assume that the learning process is a 'national' phenomenon is to accept that cognitive development has a social dimension that is distinctive to the particular nation in which learning occurs. In my view, the learning process differs signficantly even across those advanced capitalist nations that have high standards of living relative to the less developed nations of the world. Central to the process of cross-national differentiation are the strategies and structures of the business enterprises that have grown to maturation within these capitalist nations. Because capitalist societies leave so much investment decision-making in the hands of the business sector, an essential foundation for comprehending the competitive advantage of capitalist nations is a relevant theory of the competitive advantage of capitalist enterprises (for an elaboration of these views and a critical perspective on the industrial policy debate in the USA, see Lazonick, 1993).

The next section of this chapter, therefore, provides a framework for analysing the relation between investments in learning and the competitive advantage of enterprises in abstraction from the national contexts in which these investments are made and in which these enterprises operate. The following section then relates the competitive advantage of enterprises to the competitive advantage of the nations in which they are based. In the final section, I shall suggest the implications of my analysis of the role of social institutions in the learning process for the changes in international economic leadership as well as for the ability of once-dominant nations, such as Britain was in the late nineteenth century and the USA in the mid-twentieth century, to respond to the challenge of new international competitors.

THE COMPETITIVE ADVANTAGE OF ENTERPRISE

Learning and Internal Economies

A business enterprise invests in the development of human resources because the market cannot supply these resources in the quantity and quality required by the organization's product and process strategies (Porter, 1990, has termed this process 'factor creation'). By making investments that develop human resources, the business enterprise gains access to knowledge, and hence learning, that would not otherwise be available to it. Some of this knowledge becomes embodied in specialized equipment and materials, while some remains in the possession of individuals or groups of individuals.

This privileged access to knowledge in turn provides a *potential* foundation on which the business enterprise can build competitive advantage. The knowledge base is only a potential foundation for competitive advantage because the business enterprise must still effectively utilize the resources, both human and physical, that it has developed. In and of themselves, investments in productive resources burden the enterprise with fixed costs. The critical issue for competitive advantage is whether the enterprise has the organizational capability to transform the high fixed costs inherent in its investment strategy into the high-quality products at low unit costs that enable it to capture market share.

Assume that, to compete in a particular product market, an investment strategy aiming to develop human resources entails fixed costs which are higher than an investment strategy that simply makes use of 'ready-made' human resources that can be purchased on the market (for an elaboration of these arguments, see Lazonick, 1991). In Figure 10.1, *HH* depicts the cost structure that can *potentially* be generated by the high fixed cost (HFC) strategy if the human resources in which the enterprise has invested are indeed effectively developed and utilized. In contrast, *LL* depicts the cost structure of an enterprise that is competing on the basis of a low fixed cost (LFC) strategy, which requires little or no new investments in human resources by the enterprise.

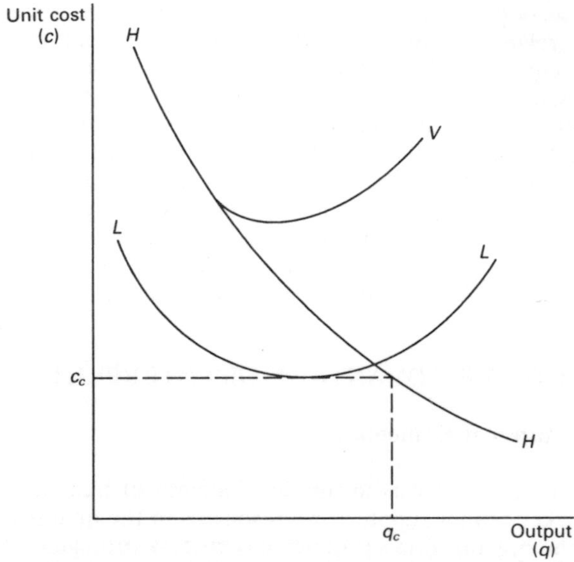

Figure 10.1 Fixed cost strategies and competitive advantage

If both *LL* and *HH* in Figure 10.1 were known cost structures, it would be rational to choose the HFC strategy that generates *HH* as long as the enterprise could supply a market demand at least as great as q_c (a quantity that I shall call 'competitive output'). Except for the case in which industry demand is simply not large enough to enable an HFC enterprise to supply output at least equal to q_c, *HH* will displace *LL* as the best-practice cost structure.

At the *outset*, however, at the particular time when the investments that entail fixed costs are made, assume that only LL is a *known* cost structure. To say that LL is known is to say that the productive capabilities of the inputs that enter into LL have already been developed and that the prices of these inputs to the enterprise have been established. Perhaps LL is the result of what was, at some previous time, an HFC strategy that rendered obsolete a cost structure based on even lower fixed costs than those in LL. But what is important to the analysis is that at the time when the HFC investments required to generate the (as yet unknown) cost structure HH are being made, LL is already an established cost structure.

In contrast to LL, at the point in time when the fixed costs inherent in what, *ex post*, becomes HH are being committed, HH is as yet unknown because the *productive capabilities* that can ultimately yield higher quality products at lower unit costs do not yet exist; these resources have to be *developed* and, once developed, *utilized*. The rationale for the HFC investment strategy is the possibility that, through the development and utilization of productive resources, *HH* will come into existence. The enterprises that make the HFC investments and bring *HH* into existence will gain competitive advantage over enterprises that continue to use traditional productive capabilities which generate *LL*.

The cost structure *HH*, depicted in Figure 10.1, is therefore the outcome of an evolutionary process. Hence the reader should keep in mind that, although the cost structures depicted in each of Figures 10.1–4 are short-run curves that exist at points in time, they have long-run histories. Indeed, in discussing the transition from one figure to the next (see below), the potential paths of this historical evolution will be outlined. As for the cost structure *HH* in Figure 10.1, the productive potential of the HFC strategy remains an uncertain possibility while the productive potential of the LFC strategy is a certainty not only at the outset when the investments that entail fixed costs are made, but also during the evolutionary process in which *HH* is being generated. Hence, during this evolutionary process, the HFC strategy may put the HFC enterprise at a competitive disadvantage if it does not develop and utilize its productive resources sufficiently to lower unit costs below c_c in Figure 10.1.

By definition, the process of bringing *HH* into existence involves *innovation* – however major or minor – because it generates quality/cost outcomes that did not previously exist. Let us call the business organization that makes these developmental investments the *innovative* enterprise. In contrast, let us call the business organization that eschews developmental investments the *adaptive* enterprise.

Of course, if, relative to known cost structures, an innovative investment strategy that required the development of resources, human or physical, were to be a *low* fixed cost strategy, fixed costs would not pose a problem for the innovative enterprise. Even at low levels of output, the innovative enterprise would have a competitive advantage. But, given factor prices and work norms (and hence a given social setting), innovation invariably involves an HFC strategy precisely because to gain competitive advantage the innovative enterprise must *develop* the productive resources that can generate higher quality products and cost-reducing processes.

Innovation generally entails higher fixed costs than existing methods of production because of not only the size but also the *duration* of the developmental investments that the innovative strategy requires. The size of the developmental investment makes fixed costs larger than is the case with *LL* because the innovative enterprise must plan and coordinate the development of a more complex specialized division of labour; these are interrelated activities not only in R&D but also in marketing and vertically related production processes. The duration of the developmental investment makes fixed costs larger than with *LL* because it takes *time* for the innovative enterprise to transform its investments into products and processes that can yield low costs per unit of output. The innovative enterprise must develop its productive resources before they can be utilized, whereas the adaptive enterprise makes use of productive resources that have already been developed, either internally or externally to the business organization itself.

For an investment of any given size, the longer the duration between the commitment of resources and the achievement of low unit costs and the consequent generation of returns, the higher the fixed costs that must ultimately be recouped. In terms of Figure 10.1, a shortening of the time from the commitment of resources to the generation of returns reduces fixed costs and, all other things being equal, thereby pushes the point of competitive output (q_c) to the left, thus making competitive advantage attainable with a smaller market share.

Central to reducing the duration component of fixed costs is the rapidity of the learning process that transforms productive resources into their developed states. Hence, rapid learning enhances the ability of the innovative enterprise to transform the competitive disadvantage inherent in its HFC strategy into the competitive advantage that comes from generating higher-quality products at lower unit costs.

What determines the rate of learning within an enterprise? The 'learning curve' concept posits that the rate of learning is a function of output; the more rapidly the enterprise achieves a large volume of output, the more rapid the rate of learning (for a survey of the learning curve literature, see Dutton, Thomas and Butler, 1984). Some economists have posited the existence of a learning curve, and have then analysed the implications of pricing strategies that sell products below current costs for the sake of gains in market share that will move the enterprise down the learning curve more rapidly than if current prices had reflected current unit costs (see Spence, 1981). Although such strategies may have the posited impact on productivity and unit costs, it is not clear why, subject to the limitations imposed by elasticity of demand for the industry's output, all enterprises in an industry would not choose a below-current-cost pricing strategy. If they did all choose such a strategy, the productivity of all the enterprises in the industry would grow more rapidly than if the constituent enterprises had priced at or above current costs, but no individual enterprise would gain competitive advantage.

Investment strategy is a much more potent determinant of the competitive advantage of enterprises than pricing strategy. Given an investment strategy, the key determinant of the rate of learning is the organizational structure of the enterprise. Even within a given national industry, the organizational structures of business enterprises differ in terms of both the productive capabilities that have been put in place and the extent to which its personnel are committed to the goals of the enterprise. These differences can have significant enterprise-specific impacts on the rate and direction of the learning process required to generate higher quality products at lower unit costs. It is those enterprises that possess superior learning capabilities – which have been elsewhere subsumed under the heading 'organizational capabilities' (Lazonick, 1990b) – that can transform HFC investment strategies from a temporary competitive disadvantage into a sustained competitive advantage.

How does organizational structure affect the rate of learning? One function of an organizational structure is to put in place an incentive system that influences the amount of effort supplied by participants in the specialized division of labour (for an elaboration, see Lazonick, 1990a). A greater supply of effort per unit of time in doing known tasks results in more rapid learning in how best to perform those tasks. In time, a producer's application of concentrated and continuous effort in the performance of particular tasks results in an internalization of the knowledge required to perform these tasks so that they become what Richard Nelson and Sidney Winter have called 'routines' (Nelson and Winter, 1982, Chs 4 and 5). In addition, and of more importance to the innovation process, a producer's application of concentrated and continuous effort may be essential to the acquisition of new knowledge that cumulates on the basis of previously acquired knowledge. Indeed, concentrated and continuous effort may be essential for achieving breakthroughs in the understanding of a problem: breakthroughs that would not have occurred had the supply of effort been less concentrated and continuous.

Therefore, the quantity and quality of effort that individual producers supply to the enterprise, and hence the rate and direction of learning that the enterprise acquires, depend critically on the organizational structure of the enterprise. But the impact of organizational structure on learning goes beyond the responses of individual producers to the incentives that the enterprise holds out. Besides requiring effort that is concentrated and continuous, and hence cumulative, learning in a modern business organization is typically collective. Learning derives not only from the efforts of individuals, but also (and more potently) from the ways in which these individual efforts of the participants in the specialized division of labour are combined. It is the organizational structure of the enterprise that determines how and to what extent the efforts of individuals coalesce to constitute a collective productive force.

Organizational capabilities enable the enterprise to affect not only the rate but also the direction of learning. Unlike the image of the trajectory of cost reductions conveyed by the learning curve literature, the learning process does not necessarily result in a continuous decline in unit costs. Indeed, it may only be with the onset of increasing costs that the enterprise becomes aware of what productivity problems must be solved and what it is that must be learned. As illustrated by the cost curve, *HV*, in Figure 10.1, an HFC strategy may be unsuccessful as internal diseconomies of scale set in before unit costs have been driven down to a level that the LFC enterprise cannot achieve (as would have been the case had *HH* rather than *HV* evolved from the HFC strategy). At any level of output, the LFC enterprise has competitive advantage. An adaptive investment strategy would appear to have won out over an innovative investment strategy.

The cost curve, *HV*, however, need not necessarily represent the final outcome of the HFC enterprise's investment strategy. The very experience of increasing costs may focus the attention of the managers of the HFC enterprise on problems in implementing the innovative strategy that (precisely because innovation is involved) could not have been foreseen at the

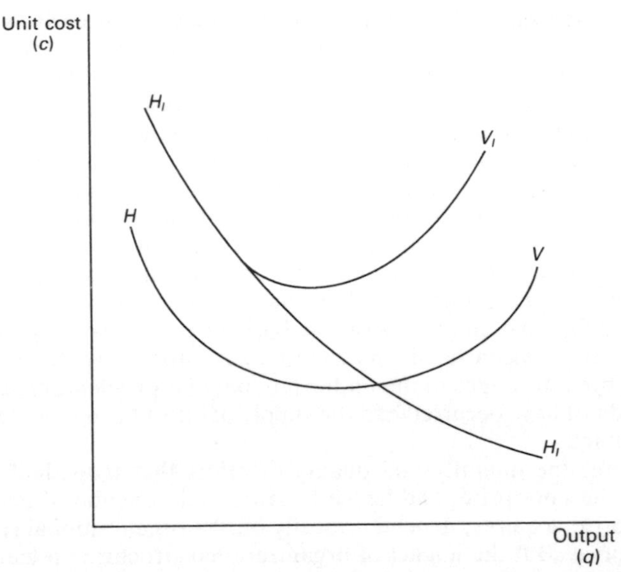

Figure 10.2 Transforming internal diseconomies into internal economies

outset. To eliminate the source of increasing costs, the HFC enterprise must make even more – and different – investments in productive resources, and incur even more fixed costs.

As depicted in Figure 10.2, the HFC enterprise that has experienced the cost curve *HV* may, through an appropriate investment strategy that complements its original investment strategy (that is, the strategy that generated *HV* in Figure 10.1) be able to 'unbend' the U-shaped cost curve and transform internal diseconomies of scale into internal economies of scale. If the HFC enterprise is successful in this strategy, the result is H_IH_I; if not, the result is H_IV_I. Because we are dealing in evolutionary processes, however, even H_IV_I might be the prelude to even more developmental expenditures that further increase fixed costs but which put in place the potential for transforming high fixed costs into high-quality products at low unit costs. Innovation is an ongoing evolutionary process in which the enterprise broadens the range of activities in which it has invested and hence the direction of learning that becomes essential to its success.

External economies and learning

As it attempts to achieve a larger extent of the market, the enterprise may experience *external* as well as internal diseconomies of scale (the distinction is, of course, due to Alfred Marshall; see Marshall, 1961, Book IV, Ch. 9;

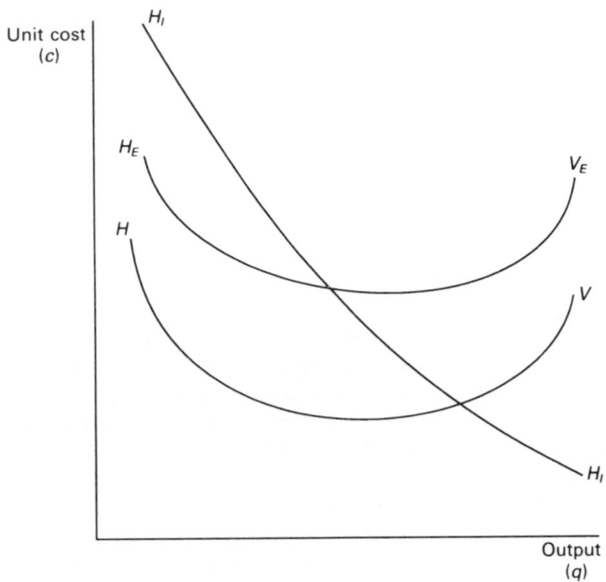

Figure 10.3 Transforming external diseconomies into internal economies

see also Lazonick, 1991, Ch. 5). Internal economies and diseconomies of scale reflect changes in *factor productivity* (physical output per unit of input) as the volume of output changes, and are determined by the organizational capability of the enterprise in producing for varying extents of its product market. In contrast, external economies and diseconomies of scale reflect changes in *factor prices*, and are determined by changes in the relation between the growth of the demand for inputs on factor markets relative to the growth of supply. Whereas internal economies and diseconomies influence the shape of the cost curve, external economies and diseconomies alter the *position* of the cost curve. Whereas internal diseconomies of scale arise because of a decline in the average quality of inputs into the enterprise's productive activities, external diseconomies of scale arise because of an increase in the market prices of inputs of a given quality. When an enterprise experiences external diseconomies, its cost curve shifts up, as depicted by H_EV_E in Figure 10.3.

As illustrated by H_IH_I in Figure 10.3, in response to the appearance of external diseconomies the enterprise might make investments that integrate into its internal organization the provision of the input that has become more expensive. For the enterprise,the goal of this investment strategy is to develop its productive resources so that it can supply itself with the input at a lower unit cost than that inherent in H_EV_E. Alternatively, the appearance of external economies might induce the firm to invest in new technologies that reduce its reliance on the input that has become more expensive. In such cases, the enterprise in effect tries to use its internal organization to overcome the constraints that market forces impose on enterprises in general. In the process, the enterprise alters the direction of learning that is critical to its success.

THE COMPETITIVE ADVANTAGE OF NATIONS

The Sources of British Competitive Advantage

If, within a national industry, internal organization can be a source of competitive advantage or disadvantage for particular business enterprises, in international competition the nation as a specific geopolitical region with a distinctive social structure and distinctive social institutions can be a source of advantage or disadvantage for those business enterprises that develop and utilize their key productive resources within its national boundaries and subject to its national norms, institutions and laws (or what might be called, from the perspective of the enterprise, its external environment).

If all enterprises within a national industry have the same cost structures and if they all operate subject to the same external environment – for example, if none attempts to transform external diseconomies into internal economies as in Figure 10.3 – then no enterprise within the national industry can gain competitive advantage over the others. Yet through the development of skill, the application of effort and external economies of scale a national industry (made up of a large number of enterprises with similar cost structures) may well be able to gain competitive advantage over other national industries that operate in other national environments.

The case of British industry exemplifies the role of the region in laying the basis for competitive advantage. In an age of imperfect machine methods, the learning of craft skills was central to generating the productive capabilities that transformed inputs into output. Although much, if not most, of the relevant learning occurred on the shop floor, this learning did not result from the investment strategies of enterprises. Instead, the training of workers with the requisite industrial skills occurred in particular industrial communities where older workers trained younger workers on the job according to standards set not by the employers in the enterprises in which production was carried out, but by the craft traditions and regulations characteristic of the region.

As a result, regions developed abundant supplies of specialized labour. Given an industrialist's choice of business (itself typically a function of his or her own specialized training in a particular region), he or she would tend to invest where labour with the necessary specialized skills was in relatively abundant supply. As a consequence, particular industries became increasingly concentrated in particular regions of Britain during the nineteenth century.

The regional concentration of specific British industries meant that employers had access to large supplies of labour with 'ready-made' skills. Given the on-the-job character of the training process and the unimportance of formal schooling for preparing the future workforce, the development of these shop-floor skills entailed relatively low fixed costs. Employers benefited from ready access to a highly productive labour force that not only kept the imperfect machinery in motion but also passed on skills to younger workers.

The fixed costs required to engage in business were also kept low by the evolution of vertically specialized and horizontally fragmented industry structures. The growth of a regionally concentrated industry facilitated the vertical specialization of constituent enterprises in narrowly defined activities, with enterprises in a particular vertical specialty relying on enterprises in other specialties to supply them with the necessary inputs and to purchase their outputs for further productive transformation and for sale downstream. The tendency towards vertical specialization was self-reinforcing because the growing availability of suppliers and buyers for intermediate products made it all the more easy for new enterprises to set up as specialists. Hence it was the entry of new enterprises rather than the growth of existing enterprises that characterized the expansion of a regionally concentrated industry. Vertically specialized industries became horizontally fragmented industries.

The evolution of industry structures characterized by regional concentration, vertical specialization and horizontal fragmentation as well as the ability of individual employers to rely on skilled workers to organize work and train younger workers on the shop floor diminished the need for business enterprises to invest in managerial personnel who could plan and coordinate the production process. The absence of managerial structures in turn reinforced the tendencies for industrial structures to become fragmented and specialized. Limited in their managerial capabilities, enterprises tended to remain under proprietary control and confined to single-plant operations.

In nineteenth-century Britain, the fixed costs of engaging in business were, therefore, relatively low. The low levels of investments in internal organization in British enterprises made it difficult for one firm to gain competitive advantage over another. The existence of craft unions, moreover, led to the standardization of labour costs across firms in a particular line of business, with increases in labour productivity being divided between higher standards of living for workers and low unit labour costs for enterprises.

While the ready availability of skilled and cooperative labour ensured high throughput within enterprises, the access of enterprises to communication and distribution networks enabled work-in-progress to flow rapidly through the vertically specialized branches of the regional industry, and ultimately to find consumers at home and abroad. As the extent of the market for a regionally concentrated industry expanded, all the enterprises in the industry experienced external economies as the fixed costs of infrastructural

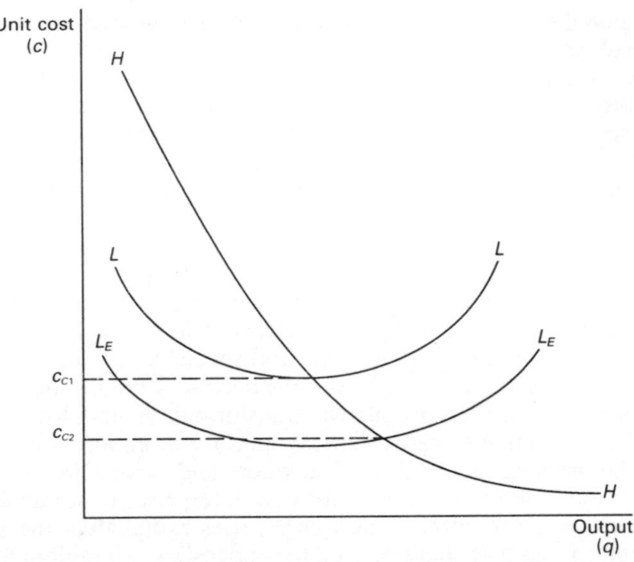

Figure 10.4 External economies as a competitive strategy

investments in communication and distribution networks were spread over a larger industry output. In effect, internal economies of scale in one sector of the regional economy led to external economies of scale for a vertically related sector. For example, with the growth of traffic, railways serving the industry would experience internal economies of scale, which would in turn be passed on to all industrial enterprises in the form of lower freight rates. The result was the downward shift in the cost curve of each industrial enterprise, as illustrated by the movement from LL to L_EL_E in Figure 10.4.

Relying more on markets than managers to coordinate industrial activity, and hence more on external economies than internal economies to lower costs over time, various British industries – iron and steel, machinery, shipbuilding and textiles – gained distinct global competitive advantages during the nineteenth century. It was the institutions of a market-coordinated proprietary capitalism, including heavy reliance on a self-reproducing supply of skilled labour to coordinate as well as execute work, that permitted British manufactures to dominate world markets in the late nineteenth century.

The Sources of US Competitive Advantage

During the first decades of the twentieth century, international industrial leadership shifted away from Britain and towards other advanced capitalist nations, particularly the USA, Germany and Japan. Accompanying the change in industrial leadership was a dramatic transformation in the institutional structure of the leading capitalist economies. The central transformation was the rise of managerial structures that, through the planned coordination of specialized divisions of labour, could generate economies internal to the enterprise. Along with the development of managerial structures came the separation of ownership from control, as the knowledge critical to running the enterprise increasingly came to reside in the heads of professional managers rather than legal owners.

The industrial corporations that generated internal economies and gained distinct and sustained competitive advantages did so by making substantial investments in the development of the capabilities of their managerial personnel. Management development took three forms. First, in the early stages of a manager's career, the employee gained expertise in a technical specialty relevant to the corporation's investment strategy and organizational structure. Second, through rotation across functional departments, those technical specialists who displayed the most skill and applied the most effort to the performance of their functions were given the opportunity to learn about other technical and organizational aspects of the operations of the enterprise, thus providing these specialists with a more general comprehension of the capabilities of the enterprise. Third, those managerial personnel who could effectively transcend their specialist backgrounds climbed up the hierarchical ladder of the corporation to positions that required more authority and responsibility in coordinating, and ultimately planning, the activities of the enterprise. In making the investments and providing the opportunities to transform the manager from a specialist into a generalist, the corporation offered the manager a 'lifetime' career path within the enterprise (see Lazonick, 1986).

The industrial corporations that were best able to offer such internal careers were those that, through attaining competitive advantage, experienced rapid and sustained growth. The ability of the dominant enterprise to utilize these scarce managerial resources that it had developed in turn contributed to its continued success. The availability of knowledgeable and committed managerial personnel made it possible for the enterprise to gain the benefits of the development and utilization of productive resources through vertical integration (see the discussion of Figure 10.2, above). The availability of these managerial resources also permitted the enterprise to expand its operations into new geographic areas, both nationally and internationally, as its investment strategy warranted.

Given the growing importance of management development within the enterprise for gaining competitive advantage in the twentieth century, to what extent was the building of managerial structures a national phenomenon? The national environment influenced both the investment strategies and organizational structures of major corporations.

The national environment influenced the investment strategies of major corporations by providing tariff protection against low-cost foreign competition during a period when, in line with the classic 'infant industry' argument, the domestic economy had yet to transform HFC strategies into unit costs that were sufficiently low to engage in open competition. Assume in Figure 10.4 that *HH* represents the cost structures of American enterprises and $L_E L_E$ represents the cost structures of British enterprises in a particular industry (say machinery manufacture) at a particular time (say 1900). With *HH* in place by 1900 the American enterprises had gained competitive advantage. But assume that, at an earlier date (say 1880), only some portion of *HH* above c_{c1} had existed because the HFC strategy adopted by American enterprises had only partially evolved, and hence had not yet been transformed into sufficiently low unit costs. In 1880 American corporations had found themselves at a competitive disadvantage in open competition with their British rivals who (consistent with the arguments made previously) had *LL* cost structures based on the utilization of traditional technologies. The role of tariff protection was to provide American corporations with the time to develop and utilize their productive resources so that the high fixed costs of their investment strategies could indeed be transformed into low unit costs and competitive advantage: that is, so that they could put the lower portion of *HH* in place (on learning and tariff protection: see David, 1975, Ch. 2).

Even as American machine makers did so, British machine makers continued to reap external economies of scale from their initial dominance of growing world markets as well as from the application of unremunerated effort by British workers in response to the growing American challenge (for a case study of cotton textiles, see Mass and Lazonick, 1990). As a result, the British cost curves shifted down from *LL* to $L_E L_E$ over the two decades, thus prolonging the ability of the British industry to remain competitive. In the end, the internal economies generated by the American enterprises gave them competitive advantage over their British rivals (and in the process overcame the need for tariff protection). But the very ability of the British enterprises to remain competitive for a time by cutting costs on the basis of their traditional organizations and technologies delayed their perceived need to embark on their own HFC strategies which could have positioned them to regain competitive advantage in the long run.

Instead, under growing competitive pressures from innovative American enterprises, the British enterprises pursued adaptive strategies of further cost-cutting on the basis of their traditional organizations and technologies. Even after the British enterprises had exhausted the possibility of cutting costs by improving their productive resources, they could reduce costs further by neglecting to replace ageing plant and equipment at the usual rate, by wringing wage concessions from immobile workers fearful of losing their jobs and by taking a lower rate of return on their invested capital, all of which would appear as reductions in factor prices and hence as downward shifts in their costs curves.

At some point (say 1930) the ability of British enterprises to cut costs on the basis of their existing productive capabilities ran out. In the absence of an innovative investment strategy, tariff protection and government subsidies only served to prolong the viability of these adaptive enterprises.

While, over the course of half a century, American enterprises had been building organizational structures to implement their HFC strategies, British enterprises had been able to adapt on the basis of their traditional organizations and technologies. As a result, British enterprises did not develop the organizational and technological capabilities required to engage in innovative investment strategies. In effect, the prolonged reliance on traditional organizational structures constrained British enterprises to rely on adaptive investment strategies, even as the viability of adaptation as a competitive response was running out.

Had British enterprises attempted to engage in innovative investment strategies in the early twentieth century, they would have had to change more than just the structures of their business organizations for, in the rise to industrial leadership of the USA, Germany and Japan, profound institutional transformations occurred external to the enterprise that complemented the building within the enterprises themselves of organizational structures to plan and coordinate the innovative investment strategies. In the USA, most prominent among the institutional changes in the national environment that supported the investment strategies of its managerial enterprises were (a) the transformation of capital markets, (b) the transformation of the educational system, and (c) the transformation of worker organizations. By briefly outlining how these institutional transformations affected the learning process in the USA, I shall indicate why the rise of US managerial capitalism to international industrial leadership during the first half of this century can be conceived as a national phenomenon.

The critical transformation in US capital markets that supported the learning process in American industry was the rise of a market in industrial securities from the 1890s (Navin and Sears, 1955). Owner-entrepreneurs who had built their firms up from new ventures to major going concerns during the last decades of the nineteenth century could now monetize their accumulation of real assets through a transfer of ownership. The old owners could now exit from their enterprises. In doing so, they left control over strategic decision-making in the hands of career managers who had earlier been recruited by the original owner-entrepreneurs as part of their (as it turned out) successful strategies to build organizational structures that would allow their enterprises to gain sustained competitive advantage (for an elaboration of these arguments, see Lazonick, 1992). The new owners were not these managers, but rather portfolio investors who lacked any detailed knowledge of the organization and technology of the enterprise. In addition, with shareholding widely distributed among many owners, these portfolio investors lacked the power to influence the investment strategy of the enterprise.

When the original owner-entrepreneur exited from the enterprise, the only productive resources that went with him or her was the 'old' learning. At the same time, his or her departure made it possible for those career managers most capable of providing leadership in charting innovative investment strategies to rise to positions of strategic decision-making. Thus the separation of ownership from control provided career managers with greater incentives to commit themselves to the goals of the enterprise. Moreover, the new possibilities for top management succession now made it possible to replace 'old' learning with 'new' learning in the strategic decision-making process.

The new learning that was relevant for industrial innovation was increasingly science-based. In the most dynamic industries, in-house development of the capabilities of scientists and engineers became central to the success of enterprises (see Chandler, 1990, Ch. 5; Mowery and Rosenberg, 1990). In making these investments in highly trained human resources, the corporations required recruits who entered their organizations with a conceptual comprehension of science and technology on which internal development programmes and on-the-job experience could build. From the late nineteenth century the American system of higher education acquired a science-based orientation that could provide the industrial scientists and engineers with the conceptual knowledge that employment in a science-based enterprise required. Moreover, through personal connections (often between university teachers and their former students working in industry), there was an ongoing exchange of information that kept the basic research in university laboratories in touch with the developmental research in corporate laboratories (see Noble, 1977; Reich, 1985; Hounshell and Smith, 1988).

However, the impact of the US system of higher education went far beyond its contribution to the science-based industries. Under the direction of the US Department of Agriculture, from the 1890s the government-supported land-grant college system experimented with new technologies for agriculture, thus performing a function that the farmers could never have undertaken themselves. In addition, through agricultural extension services, the land-grant colleges helped to ensure that farmers would become aware of the existence of successful new technologies and be instructed in the use of them (Ferleger and Lazonick, 1992).

Beyond developing technologies or technologists, after the turn of the century the system of higher education was increasingly called on to provide the general cognitive and behavioural formation that those destined for line as well as staff positions would require to function within a bureaucratic organization. By the 1920s the American system of higher education had taken its present form, providing the pre-employment foundation for managerial development within the enterprise. Educated recruits could be expected to have the cognitive capabilities necessary for acquiring industry-specific technical knowledge, as well as the behavioural characteristics required to interact with others within the organizational context and respond positively to organizational incentives (Lazonick, 1986).

Not all corporate employees benefited from such education and internal training. In the late nineteenth century the managerial quest to use mass production technologies to achieve high levels of throughput and low unit costs confronted the attempts by workers to exercise craft control over the workplace. The result was a growing conflict between management and labour. As the corporations became more concentrated and grew stronger around the turn of the century, workers built their craft unions. The American Federation of Labor, founded in 1886, experienced its greatest growth in membership around the turn of the century while American industrial corporations were going through the Great Merger Movement. If there was one objective that united American industrial corporations during the first three decades of this century, it was to keep their enterprises union-free. Indeed, these decades have become known as 'the non-union era' in American labour history (see Brody, 1980, Ch. 1).

How did the American corporations contain the organized labour movement in the early decades of the twentieth century, and what are the implications of this victory of corporate management over the labour movement for the competitive advantage of the American enterprise? An adaptive response was to transform key craft workers, generally paid by the piece, into salaried members of the managerial structure, thus securing their commitment to the goals of the enterprise (Montgomery, 1987). A more innovative response was to do away with the need for craft labour by adopting skill-displacing technologies (see Lazonick, 1989).

In so doing, American corporations continued a trend that had begun in the first half of the nineteenth century. The national environment of the USA was one in which skilled labour was highly mobile between enterprises and industries. Highly mobile skilled labour can command high wages and cannot be compelled to deliver high levels of effort. The adoption of skill-displacing technologies freed management from reliance on this expensive and unreliable 'variable' factor of production. In its place (as indicated by the shift from *HV* to $H_I H_I$ in Figure 10.2), management now had a fixed factor over which it could exercise more control.

How did American managers exercise this control? How did they transform the high fixed costs of the new technologies into low unit costs? To tend the new technologies, corporate managers substituted unskilled operatives (mainly immigrants from southern and eastern Europe) for the skilled craft workers (mostly of British and German origin). They then increased the complements of shop-floor supervisory labour to monitor the effort of these workers. But, especially in periods of boom when labour markets were tight and when employers most wanted to achieve high levels of throughput, overbearing supervisors could be counterproductive. Knowing that their labour was in short supply, alienated workers might engage in slowdowns, or even in the sabotage of expensive machinery and materials, rather than feel compelled to supply more effort (Lazonick, 1990a, Ch. 7).

Ultimately, it was not simply close supervision that transformed high fixed costs into low unit costs but the offer to shop-floor workers of 'good jobs': employment opportunities that could not easily be found elsewhere in the economy (see Jacoby, 1985). The enterprises that were best able to make these superior employment offers were those that, through innovative investment strategies and appropriate organizational structures, had already gained dominant positions in their industries. Indeed, by the 1920s, those American enterprises which, through the superior development and utilization of their productive resources, had gained distinct competitive advantages, were able to cumulate those advantages by inducing their blue-collar workforce to supply high levels of effort in exchange for the promise of relatively high pay and employment security. These high levels of effort were critical for achieving high levels of throughput on mass production machine technologies, thus helping to transform the high fixed costs of the investments in these technologies into low unit costs.

Yet in recent decades the very strategy of investing in technologies and organizations that permit the utilization of deskilled shop-floor labour has proved to be the Achilles heel of American industry in international competition. The Japanese in particular have been able to gain competitive advantage by developing the skills not only of white-collar employees within the managerial structure but also of blue-collar employees on the shop floor.

The strength of American industry in the post-Second World War decades lay in the development of skills of those salaried personnel who were deemed to be part of management: an investment strategy that, as we have seen, dated back to the nineteenth century. It was through the collective learning of these personnel that American enterprises retained unmatched organizational capability for product and process innovation.

The advent of mass-production unionism from the late 1930s gave rise to seniority provisions that compelled the major American industrial corporations to grant blue-collar workers long-term employment security. But these workers were still deemed to be 'hourly employees'. As such, corporate employers fought hard and successfully to ensure that these blue-collar workers had no formal control over work organization on the shop floor and no formal right to influence, or even question, management's investment strategy. In particular, the coming of mass-production unionism did not result in investment strategies to upgrade the skills of blue-collar workers. On the contrary, as workers gained more collective power, American managers became all the more determined to take, and keep, skills off the shop floor.

Sources of Japanese Competitive Advantage

A prime source of Japanese competitive advantage derives from the strategy of investing in the skills of shop-floor employees. This strategy contrasts with that of enterprises in Britain where, for lack of appropriate management structures, employers left it up to workers to develop skills that could be used on the shop floor. British workers developed those skills that furthered their own particular craft interests. The Japanese strategy also contrasts with that of enterprises in the USA where, because of the historical attempt by workers to maintain craft control, employers in possession of appropriate management structures became obsessed with taking skills off the shop floor. The unwillingness of US corporate management to invest in the development of human resources on the shop floor has ensured a clear-cut organizational segmentation between management and labour within American enterprises, and this segmentation can pose a formidable barrier to effective planned coordination of the activities of the enterprise (Lazonick, 1990a; see also Florida and Kenney, 1990).

In recent years, the Japanese strategy has provided blue-collar workers with skills that are relevant to the successful operation of just-in-time inventory systems, statistical quality control and flexible manufacturing technologies. By combining the skills of shop-floor workers with mass production methods, Japanese enterprises have been able to achieve high levels of throughput while attaining high standards of product quality and reducing the fixed costs of inventories to a minimum. More than that, the combination of shop-floor skills and machines has given them the ability to engage in 'flexible mass production': the use of the same plant and equipment to produce a variety of differentiated products without having to resort to low-throughput, batch-production methods (see Lazonick, 1990a, Ch. 9). What some observers have called 'lean production' (Krafcik, 1988) depends critically on developing the skills of shop-floor workers.

What accounts for the willingness of Japanese enterprises to make investments in shop-floor workers that US enterprises have generally been unwilling, and British enterprises generally unable, to make? As argued in *Competitive Advantage on the Shop Floor* (Lazonick, 1990a), the distinguishing feature of the history of Japanese labour–management relations was the absence in Japan of a significant supply of craft workers during the period in which industrial capitalism emerged as a viable economic system (Lazonick, 1990a, Ch. 9; for a detailed historical account, see Gordon, 1985). When Japan began its process of rapid industrial development in the last decades of the nineteenth century, enterprises could not draw on an already developed supply of industrial skills as was the case in nineteenth-century Britain and the USA. Japanese employers could not, therefore, adopt the British strategy of relying on an abundant supply of craft workers to plan and coordinate work on the shop floor, and neither, as in the American case, did Japanese employers have to confront the problem of introducing high-throughout technologies in opposition to groups of craft workers intent on maintaining their traditional craft prerogatives. Instead the shop-floor problem for Japanese enterprises was one of developing industrial skills, and then maintaining access to them.

Long before the advent of 'permanent employment' in the decades after 1945, Japanese enterprises had sought to overcome the scarcity of industrial skills by investing in the capabilities of their shop-floor workers, and particularly male workers who could be expected to spend their working lives with the company. Having made the investment, these enterprises then structured internal employment incentives to ensure that they maintained access to these human resources over the long term.

The Japanese strategy for developing and utilizing the skills of shop-floor workers has been possible only because Japanese enterprises made the complementary investments in managerial organizations that could plan and coordinate specialized divisions of labour. Early in the twentieth century, Japanese enterprises were already engaged in the widespread recruitment of college graduates (many of whom had to be lured away from prestigious government posts) to serve as line and staff personnel (Yonekawa, 1984; Daito, 1986; Morikawa, 1989; Uchida, 1991). As a result of this organization building, the Japanese have been better able than the Americans or British to ensure that the activities of shop-floor workers are integral to the overall strategies and structures of their enterprises.

Supporting these investments in management and workers has been a national environment in which government policy has played a major role (see Johnson, 1982; McCraw, 1985; Freeman, 1987; Best, 1990, Ch. 6). As was the case historically in the USA, the Japanese state has protected the home market to permit business organizations to develop and utilize their productive resources to the point where they could attain competitive advantage in open international competition. But the Japanese state has gone further: it has maintained a stable macroeconomic environment characterized by high levels of employment and a relatively equal distribution of income across sectors, thus enlarging the extent of the Japanese market for manufactured goods. It has increased incentives for consumers and businesses to purchase goods (for example, computers and televisions) that embody state-of-the-art technologies. It has limited the number of firms competing in major manufacturing industries, thus creating incentives for these firms to incur the high fixed costs necessary to attain competitive advantage. It has promoted cooperative research and development among major Japanese competitors, while ensuring manufacturing corporations with access to long-term finance at low rates. And the Japanese state has provided industry with a highly educated labour force to fill blue-collar, white-collar and management positions. But however important the role of the Japanese state in shaping an environment conducive to economic development, the formulation of investment strategies and the building of organizational structures to carry them out has been entrusted to private-sector enterprises (for industry studies on business-government relations, see Collis, 1988; Anchordoguy, 1989; Fransman, 1990).

Until the 1980s it was tempting for foreign competitors to assume that the main source of competitive advantage of these Japanese enterprises was not their ability to develop and utilize productive resources but instead their ability to pay their workers low wages and work them long hours. These sources of competitive advantage have played a role in Japan. But, as foreign visitors to Japan learned when they visited cotton mills in the late 1920s and early 1930s (see the references in Mass and Lazonick, 1990) and car plants in the late 1970s and early 1980s (see, for example, Abernathy, Clark and Kantrow 1983), the belief that Japan's competitive advantage lay simply in an exploited labour force missed the importance of continuous, cumulative and collective learning in generating economic development that was both substantial and sustained.

LEARNING, INSTITUTIONS AND 'CONVERGENCE'

As has been argued elsewhere, the institutional foundations of Japan's rise to international industrial leadership in the twentieth century do not constitute a wholly new model of advanced capitalist development, but more a thoroughgoing elaboration of the private-sector institutions for planning and coordinating specialized divisions of labour that carried American managerial capitalism to its pre-eminent position in the first half of this century (Lazonick, 1991, Ch. 1). The history of the change in industrial leadership from the USA to Japan over the past few decades does demonstrate, however, that the social structure of economic institutions exerts an influence – and, I would argue, a preponderant influence – on the rate and direction of learning. Over time and across leading capitalist economies, the learning that lays the foundation for industrial leadership is increasingly continuous, cumulative and collective. For such learning to occur it is necessary for the institutions which develop and utilize productive resources, and hence generate economic development, to plan and coordinate specialized divisions of labour that are increasingly complex and costly. Through planned coordination, these institutions must ensure not only the appropriate collective cognitive development of human resources but also appropriate behavioural responses on the part of the participants in the specialized division of labour in whom investments have been made, for it is only through the appropriate behavioural responses that the investments in cognitive development are utilized sufficiently to transform high fixed costs of investments in innovative learning into low unit costs.

In capitalist economies, the central institutions for developing and utilizing productive resources are business enterprises: enterprises which must capture sufficient market shares to lower unit costs and generate the revenues to justify their continued existence. Yet, as has been argued in this chapter, the strategies and structures of these business enterprises are influenced by the social environments in which they acquire productive resources and in which they seek to generate revenues. Among other things, this social environment – which I have identified as the 'national' environment – determines the quality of human resources available to business enterprises to develop, as well as the financial terms on which the requisite investments in productive resources must be made.

The relation between business enterprises and their national environment can, however, be reciprocal. By the very development and utilization of productive resources, the strategies and structures of business enterprises can also shape elements of the national environment such as educational systems, financial systems and class structure (Lazonick, 1986; Lazonick, 1992, Lazonick, 1990c). Indeed, it is when business enterprises are most successful in developing and utilizing productive resources that the strategies and structures internal to these enterprises have the greatest influence on the transformation of the social system as a whole.

The perspective outlined here has important implications for understanding why changes in international industrial leadership occur, and the difficulties that former leaders have in responding to new competitive challenges. Recently, economists have raised these issues in exploring the validity of 'the convergence hypothesis' (see Baumol, Blackman and Wolff, 1989; see also Abramowitz, 1989, Chs 1 and 7). The convergence hypothesis argues that those nations which have made fundamental public investments in the education of their people – nations that belong to what Baumol, Blackman and Wolff (1989, Ch. 9) call the 'convergence club' – have the capability of borrowing new technology from technological leaders, which in turn permits them to increase their rates of productivity growth so that their levels of productivity converge on that of the leaders (Baumol, Blackman and Wolff, 1989, Ch. 5).

The convergence hypothesis has the virtue of focusing attention on the roles of education and technological change as critical determinants of long-run industrial competitiveness. But the institutional perspective on learning and competitive advantage that is offered here suggests a number of weaknesses in the convergence hypothesis. In concluding this chapter, my purpose is simply to mention these weaknesses as a prelude to further research.

First, it is argued that a sound educational system provides the foundation for followers to adopt foreign technology, thus permitting their rates of productivity to converge on that of the leader. In long-run historical perspective, however, it must be recognized that the relation between public investments in education and industrial leadership is a twentieth-century phenomenon. In the nineteenth century, Britain developed into the workshop of the world with little investment in, or assistance from, a system of public education. In contrast, in the twentieth century the development of a system of public education has become critical for successful economic performance. Yet, particularly in the USA, the form and content of the system of public education responded to the changing labour force requirements of business enterprises, and these requirements were determined by the strategies and structures of these enterprises (Bowles and Gintis, 1976; Noble, 1977; Lazonick, 1986). In all the advanced capitalist economies of this century, the willingness of the state to invest in an educational system has been of vital importance to the growth of productivity. But the rate and direction of such public investments have been profoundly shaped by the rate and direction of business investments. Indeed, part of Britain's economic problem in the twentieth century has been the failure of business enterprises to make widespread and effective investments in organizational and technological innovation that could in turn have placed enough pressure on, and provided enough resources to, the educational system to transform it to serve the human-resource needs of business (see Lazonick, 1986).

Second, the convergence hypothesis overestimates the degree to which technology can be adapted from one national environment to another. Insofar as the utilization of technology requires complementary human inputs with specific cognitive capabilities and behavioural responses, the transferred technology will have to be developed in the new national environment before it can be utilized there (for a case study of textile technology that compares Britain and Japan, see Mass and Lazonick, 1990). As a result, when 'transferred' technology is ultimately developed so that it can be productively utilized in a new national environment, it is in effect a new technology. Indeed, in terms of its impact on productivity, it may be a superior technology. Historians of technology now recognize the importance of the social determinants of technological change within particular national environments (see MacKenzie and Wajcman, 1985; Bijker, Hughes and Pinch, 1987). The analysis of the social determinants of technological change must also be put in comparative, cross-national perspective.

Third, because social organization is so important to the development and utilization of productive resources, former technological leaders may have problems responding to competitive challenges from national industries that, through the dynamic interaction of organizational structure and technological change, have generated rapid rates of productivity growth. With institutions in place that were appropriate for the development and utilization of what have by now become traditional technologies, the former technological leaders may be incapable of increasing their rates of productivity growth by simply 'borrowing back' technology from the rising competitors. They may also lack the organizational capabilities to develop and utilize superior technologies. Instead, they may – as was done first by Britain and more recently by the USA – pursue the alternative strategy of adapting on the basis of their traditional organizations and technologies (Lazonick, 1991, Ch. 1). By avoiding HFC investments in organizations and technologies that would have generated returns only in the future, this adaptive strategy may augment rates of productivity growth in the short run while making it impossible to sustain these rates of growth in the long run. It is important, as the proponents of the convergence hypothesis argue, to focus the analysis of economic performance on the long run; but future research must also recognize that it is the dynamic interaction between organization and technology that determines whether the long run results in competitive success or relative decline.

References

Abernathy, William J., Clark, Kim B. and Kantrow, Alan M. (1983) *Industrial Renaissance: Producing a Competitive Future for America* (New York: Basic Books).

Abramowitz, Moses (1989) *Thinking about Growth* (Cambridge University Press).

Anchordoguy, Marie (1989) *Computers, Inc.: Japan's Growing Challenge to IBM* (Cambridge, Mass.: Harvard Business School Press).

Arrow, Kenneth (1962) 'The Economic Implications of Learning by Doing', *Review of Economic Studies*, 29 (June).

Baumol, William J., Blackman, Sue Anne Batey and Wolff, Edward N. (1989) *Productivity and American Leadership: The Long View* (Cambridge, Mass.: MIT Press).

Becker, Gary (1968) 'Investment in On-the-Job Training', in M. Blaug (ed.), *The Economics of Education* (Harmondsworth: Penguin), pp. 183–207; excerpted from Gary Becker, *Human Capital* (New York: Columbia University Press, 1964), pp. 7–29.

Best, Michael (1990) *The New Competition: Institutions of Industrial Restructuring* (Cambridge, Mass.: Harvard University Press).

Bijker, Wiebe E., Hughes, Thomas P. and Pinch, Trevor J. (eds) (1987) *The Social Construction of Technological Systems* (Cambridge, Mass.: MIT Press).

Bowles, Samuel and Gintis, Herbert (1976) *Schooling in Capitalist America* (New York: Basic Books).

Brody, David (1980) *Workers in Industrial America* (N.Y.: Oxford University Press).

Chandler, Alfred D. Jr (1990) *Scale and Scope: The Dynamics of Industrial Capitalism* (Cambridge, Mass.: Belknap Press).

Collis, David J. (1988) 'The Machine Tool Industry and Industrial Policy, 1955–82', in A. Michael Spence and Heather A. Hazard (eds), *International Competitiveness* (Cambridge, MA.: Ballinger), pp. 75–114.

Daito, E. (1986) 'Recruitment and Training of Middle Managers in Japan, 1900–1930', in Kesaji Kobayashi and Hidemasa Morikawa (eds), *Development of Managerial Enterprise* (Tokyo: Tokyo University Press), pp. 151–79.

David, Paul (1975) *Technical Choice, Innovation, and Economic Growth* (Cambridge University Press).

Dutton, J.M., Thomas, A. and Butler, J.E. (1984) 'The History of Progress Functions as a Managerial Technology', *Business History Review*, 58 (Summer).

Ferleger, Louis; and Lazonick, William (1992) 'The Managerial Revolution and the Developmental State: The Case of U.S. Agriculture', photocopy, Harvard University.

Florida, Richard and Kenney, Martin (1990) *The Breakthrough Illusion: Corporate America's Failure to Move from Innovation to Mass Production* (New York: Basic Books).

Fransman, Martin (1990) *The Market and Beyond: Co-operation and Competition in Information Technology in the Japanese System* (Cambridge University Press).

Freeman, Christopher (1987) *Technology Policy and Economic Performance: Lessons from Japan* (London: Frances Pinter).

Gordon, Andrew (1985) *The Evolution of Labor Relations in Japan: Heavy Industry, 1853–1955* (Cambridge, Mass.: Harvard University Press).

Gospel, Howard (ed.) (1991) *Industrial Training and Technological Innovation: A Comparative and Historical Perspective* (London: Routledge & Kegan Paul).

Hounshell, David and Smith, John K. Jr (1988) *Science and Corporate Strategy: Du Pont R&D, 1902–1980* (Cambridge University Press).

Jacoby, Sanford (1985) *Employing Bureaucracy: Managers, Unions, and the Transformation of Work in American Industry, 1900–1945* (New York: Columbia University Press).

Johnson, Chalmers (1982) *MITI and the Japanese Miracle: The Growth of Industrial Policy, 1925–1975* (Stanford, CA: Stanford University Press).

Krafcik, John F. (1988) 'The Triumph of the Lean Production System', *Sloan Management Review*, 30 (Autumn).

Lazonick, William (1986) 'Strategy, Structure, and Management Development in the United States and Britain', in Kesaji Kobayashi and Hidemasa Morikawa (eds), *Development of Managerial Enterprise* (University of Tokyo Press), pp. 101–46.

—— (1989) 'The Breaking of the American Working Class', *Reviews in American History*, 17 (June).

—— (1990a) *Competitive Advantage on the Shop Floor* (Cambridge, Mass.: Harvard University Press).

—— (1990b) 'Organizational Capabilities in American Industry: The Rise and Decline of Managerial Capitalism', *Business and Economic History*, second series, 19; a slightly different version of this essay appears in Gospel (1991), Ch. 10.

—— (1990c) 'Organizational Integration in Three Industrial Revolutions', in Arnold

Heertje and Mark Perlman (eds), *Evolving Technology and Market Structure: Studies in Schumpeterian Economics* (University of Michigan Press).

—— (1991) *Business Organization and the Myth of the Market Economy* (Cambridge University Press).

—— (1992) 'Controlling the Market for Corporate Control: The Historical Significance of Managerial Capitalism', *Industrial and Corporate Change*, 1.

—— (1993) 'Social Organization and Technological Change', in William Baumol, Richard Nelson, and Edward Wolff (eds), *International Convergence of Productivity with some Evidence from History* (New York: Oxford University Press).

—— and Brush, Thomas (1985) 'The "Horndal Effect" in Early U.S. Manufacturing', *Explorations in Economic History*, 22 (February).

MacKenzie, Donald and Wajcman, Judith (eds) (1985) *The Social Shaping of Technology* (Milton Keynes: Open University Press).

Marshall, Alfred (1961) *Principles of Economics*, 9th (Variorum) ed. (London: Macmillan), 1 (text).

Mass, William and Lazonick, William (1990) 'The British Cotton Industry and International Competitive Advantage: The State of the Debates', *Business History*, 32 (October).

McCraw, Thomas K. (ed.) (1985) *America versus Japan* (Boston, Mass.: Harvard Business School Press).

Montgomery, David (1987) *The Fall of the House of Labor* (Cambridge University Press).

Morikawa, Hidemasa (1989) 'The Increasing Power of Salaried Managers in Japan's Large Corporations', in William D. Wray (ed.), *Managing Industrial Enterprise: Cases from Japan's Prewar Experience* (Cambridge, Mass.: Council on East Asian Studies), pp. 27–51.

Mowery, David and Rosenberg, Nathan (1990) *Technology and the Pursuit of Economic Growth* (Cambridge University Press).

Navin, Thomas and Sears, Marion (1955) 'The Rise of a Market for Industrial Securities, 1887–1902', *Business History Review*, 24 (June).

Nelson, Richard R. and Winter, Sidney G. (1982) *An Evolutionary Theory of Economic Change* (Cambridge, Mass.: Belknap Press).

Noble, David (1977) *America by Design: Science, Technology, and the Rise of Corporate Capitalism* (New York: Oxford University Press).

Porter, Michael (1990) *The Competitive Advantage of Nations* (New York: Free Press).

Reich, Leonard (1985) *The Making of American Industrial Research* (Cambridge University Press).

Schultz, Theodore (1968) 'Investment in Human Capital', in Mark Blaug (ed.), *The Economics of Education* (Harmondsworth: Penguin), pp. 13–33; originally published in *American Economic Review*, 52 (1961).

Solow, Robert (1962) 'Technical Progress, Capital Formation, and Economic Growth', *American Economic Review*, 52 (May).

Spence, A. Michael (1981) 'The Learning Curve and Competition', *Bell Journal of Economics*, 12 (January).

Uchida, Hoshimi (1991) 'Japanese Technical Manpower in Industry, 1880–1930: A Quantitative Survey', in Gospel (1991), Ch. 5.

Yonekawa, Shin'ichi (1984) 'University Graduates in Japanese Enterprise before the Second World War', *Business History*, 26 (July).

Part IV Technological Development and Economic Transformation

The interaction between institutional innovation and technological change, a theme of many earlier chapters, characterizes whole economies as well as sectors. As the following essays argue, technological change follows a socio-technical sequence in which the institutional requirements of industrialization evolve as technology advances.

The interlinked evolution of science, technology and economy poses some of the most challenging questions in the domain of technological change. Taking the long view of the historian, William Parker argues that technology, science and the economy have their own histories built around their own institutions, practitioners and communications media. But these histories are interrelated, and they have become more so as the economy has evolved. In the Industrial Revolution, the economy simply provided an opportunity for technological change; pre-industrial development had created a needed scale of markets and density of communication networks. The nascent industrial economy then gave rise to large-scale industrial firms and professional engineers. These in turn tied technology and science ever more closely to the needs of the modern economy.

The transition to an economy of managerial firms with organized innovation systems is hardly automatic. Edward Nell conceives it to be a transformation of the growth process itself, involving both technological and institutional changes. Nell identifies two types of firm corresponding roughly to two stages in the history of the economy. One employs a team of skilled workers who use non-transferable knowledge to produce in small batches. The other mass produces standardized goods using more generic technological knowledge. Each type of firm can develop techniques: however, the transition to mass production requires such basic changes in scale, financing, marketing and knowledge that institutional change outside the industrial sector – such as new government initiatives, the formation of engineering schools and new banking systems – may first be necessary. Once established, mass production companies, within a network of capital goods firms, use competitive advantages to sustain the transformation. Moreover, the transition changes the typical cost structure of firms, exacerbating cyclical instability. Mass production firms thus require regular government intervention.

There are different routes to mass production. Alice H. Amsden and Takashi Hikino identify an important path for late industrializers. Since they entered the technology race at a different historical stage from earlier industrializers, countries such as Japan and South Korea have followed a distinct path to industrialization. Prevented from competing at the technological frontier and recognizing that low-skill, labour-intensive products had little growth potential, these countries entered 'mid-tech' industries. They could succeed because they came to excel at two learning processes: learning from technology purchased on the world market and learning how to improve the production process. Government did its part by protecting the companies with subsidies, procurement arrangements and import controls as long as the firms made progress towards agreed-upon goals. Profitable products then provided the finance and knowledge for diversification into new product lines.

11 History and its Lessons

William N. Parker

Mr Dooley, the Chicago newspaper column philosopher in the years around 1900, read in the paper one day a statement from a college professor about the state of the world. He commented to his friend, Mr Hennessy, 'Tis a grand thing to be a college professor.' Mr Hennessy replied, 'Not much to do.' 'No,' mused Mr Dooley, 'not much to do, but a *great deal to say*.'

That comment has a point to which I wish to revert at the end of this chapter. What is it that history and economics professors can do on a grand topic like technological change when most of us cannot even drive in a nail straight? Yet perhaps we may say that although we are not technologists, we have learned something about the other subject of these chapters, the human processes of learning: learning by doing, learning by reading, even learning by talking, and surely learning by teaching. In this aspect of the subject of this volume, every scholar has a professional interest and capability.

The conference from which this book grew began, appropriately enough, with reflections on technological change from three elderly (more politely, senior) scholars, each residing in a separate room of that rambling structure called history. Alfred D. Chandler (see Chapter 3) spoke as a historian of American and European business and business forms, notably of the large, multi-divisional firm now pandemic in the economic life of Western capitalism.[1] Thomas P. Hughes, as a historian of technological systems, based his analysis on his notable studies of systems of power generation and distribution, and on all such systems as both the intellectual underpinning, and the social and economic accompaniment of historical change.[2] As a professed economic historian, I considered technology from the view of one who examines the changes in wealth in human societies and the evanescent forms of social life by which such wealth is produced and distributed.[3] Noting how each of us (hyphenated historians) saw a technological change from a particular view – a view which we had taught and reiterated over the years – an observer might have recalled the famous tale of the three blind men and the elephant, or perhaps the story known to the *cognoscenti* of the joke book as 'The Elephant and the Polish Problem'.

Technological change is indeed an elephant in the kingdom of history: an animal too large to be missed and too curious, complex and multiform to be brought into even the capacious pen of historical narrative. The subject requires not a hyphenated but a general historian, sensitive to the human psyche in one of its most creative manifestations. But it is to the study of human society that history's great generalists have been attracted, rather than to the obscure and slow evolution of our relation to physical nature.

Real historians deal with leaders, opinions, influences, movements and the structures based on force, power or value that people create for themselves and one another. Usually too, the historian has, no matter how deeply hidden, an ideological or personal psychological axe to grind. He or she is a humanist; 'good' and 'bad', 'right' and 'wrong', are prominent terms in the historian's vocabulary, although covered by mountains of facts and 'woven' into narratives to give the impression of cause and effect. Such a one will tend to think of the world of nature and physical forces as the neutral setting for human struggles, and share the viewpoint that Goethe put into the mouth of Mephistopheles:

> Von Sonn' und Welten hab' Ich nichts zu sagen
> Ich sehe nur wie sich die Menschen plagen.
>
> (Of suns and worlds I nothing have to tell,
> I see but how men make their lives a Hell.)

David Landes, who went on to become a notable historian of clockwork mechanisms,[4] when writing earlier his great work on the industrial history of the nineteenth century, saw it as a race between entrepreneurial groups of different nationalities, and referred to the technological details as 'anodyne'. To write with penetration and sympathy about technological change, one must become part-technologist by means of a certain 'willing suspension of disbelief'. And when one does so, a feeling for the human scene begins to vanish from the mind. The historians of 'man, the inventor', and those of 'man, the high-citied', as A.P. Usher, himself a notable technological historian, citing Sophocles, distinguished them, are polar types.[5] We have on the highest authority that 'ye cannot serve God and Mammon'. Nevertheless, it is one's Fate, in dilemmas like this, to make the effort.

History writing, even on the history of technology, has, of course, its own technology. The traditional and, in many ways, the most satisfying one is the prose narrative. We like to say that history tells a story, and although no one can define just what a story is, we recognize and feel its thrill when we see it. It is not my purpose here to try to do that: rather it is to suggest more prosaically the outlines of the elements which such a story must contain. The mode of expression is neither the prose paragraph nor the mathematical model. The effort to be clear on an incredibly tangled subject is attacked by diagrams composed of boxes, arrows and suggestive words.

In these diagrams, technological change is taken as a thread throughout two centuries, set in its relevant, evolving historical fabric. Two other such thread-histories, definably separate from technology but closely intertwined with it nonetheless, are seen to be particularly relevant. One – the history of the economy – has developed in symbiotic relationship to technology over the past two centuries, and appears to be moving ever closer to it in the past 50 years: the economy taken, that is, as the growing set of producing, distributing and trading units in which the ever-changing body of economic values and magnitudes, of inputs and outputs measured in the altering set of prices, is brought into being. Technology's other companion on the trip through the centuries is science, a creation not of the hands, but of the intellect, developing by observation, generalization, the ordering of elements, and the construction of the models to explain (predict?) the behaviour of the forces and substances of the natural world. Here, then, are three threads to be stripped off, as it were, from the cloth of history, and hung up to dry, and the question is asked: what were their interconnections with one another, and how far can their separate records, in timing, in rate and in composition, be explained in terms of themselves alone and in terms of the constraints and opportunities offered to human individual and social action by the separate histories of either one, or both of the other two?

In a hyphenated-historian's structure of thought, each of these words – science, technology and economy – is considered as signifying a discernible historical entity itself; a sequence, a 'thread' in history's fabric as it comes off the loom of time. All such partial histories have three dimensions: (a) a product, or stream of products, as recorded by measurements and stated in words descriptive of the constituent subproducts, or items; (b) an evolving human social organization by which a portion of each human generation's energies, time and skills is combined in the creation and diffusion of the scientific ideas, the technical devices and formulae, the values of economic goods and services, which the records respectively display; (c) an underlying culture – intellectual/scientific, technical/artistic, or commercial – by which successive generations of participants are trained and initiated along these paths. All such partial cultures form, of course, components of the general, common culture of the larger social group, whose alterations through time create history: *histoire totale* (total history).

To sketch some important issues in the history of technology, in relation to science and the economy, the diagrams (Figures 11.1, 11.1a, 11.1b and 11.1c) presented here assume the existence of the *explicandum*: the factual record of scientific achievement, the stream of new technology, and – for the economy – the historical record of production, of output, factor inputs and shifting prices by which the whole is assessed.

Still the diagrams are a bit eclectic as among all three of these dimensions or 'levels': (a) the record, (b) the organization, (c) the underlying social culture. Some of the words used – 'state and private', 'workshops and large firms' – refer not only to (b), the organizational form in these activities, but would also appear in discussions of (c), institutions and features of the underlying culture. On the other hand, the references to the focus of scientific achievement – 'mechanics', 'chemistry', 'atomic physics', 'biology' – refer to features of (a), the record of science, but are basic to an understanding of the influence of scientific work on the record of technology and the economy, which the human actors in those two columns of Figure 11.1 – those two sectors of the social history – produce. Like the specific historical records, not much will also be said about (c), the common culture, the great muddy bath in which all have their feet a-soaking. The focus here will be on the social organization of the acculturated individuals pursuing the activity and, in so doing, creating its record. A portion of that social organization is thought of as its 'sponsorship' where, in the scientific and technological

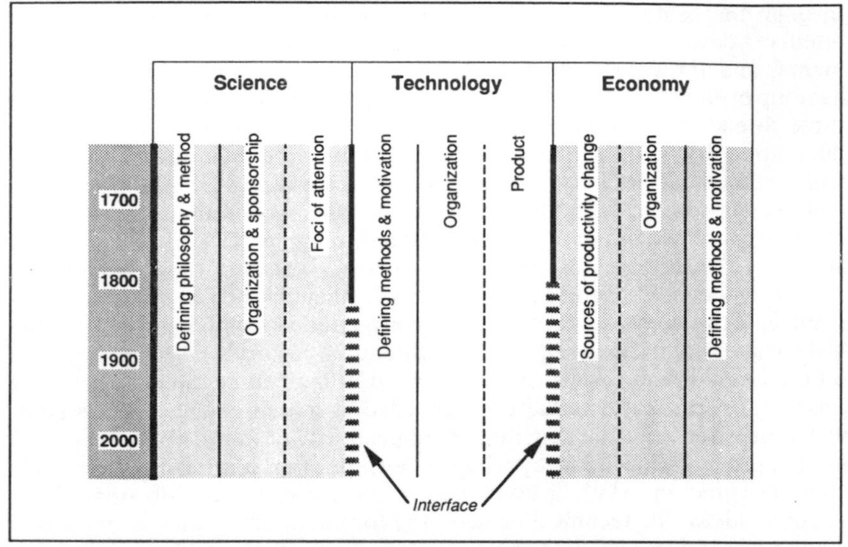

Figure 11.1 Science, technology and economy, 1700–2000: an overview

enterprises, the prime motivation and objective is not specifically a well-calculated entrepreneurial profit.

To examine any one of these diagrams and elaborate it in detail is the labour of an afternoon, not of a few pages among the thoughtful and well-researched contributions to a volume. I shall limit my comments on what the diagrams display (the 'stories' they tell, when pieced together, or overlaid across the two centuries on one another) to a few points that seem to me most significant and best displayed by this technique of exposition.

THE PRIORITY OF SCIENCE

Conceptually, Figure 11.1, which shows the three parallel 'threads', should be bent back so as to form the continuous surface of a cylinder. A body of connection and influence exists between economy and science which, on a two-dimensioned page, is sacrificed to the demands of plane geometry. The implication is that this connection has not been (until, as we shall see, in the past 50 years) very close or strong.[6] Instead the arrows run largely from left to right across the page. This is the essence of an idealistic, non-materialist theory of history. Science, with its well-developed set of values and standards, deploys itself across the range of natural phenomena – forces and substances – creating one branch after another in an unfolding array. Technology, starting first without close intimacy or interchange with sci-

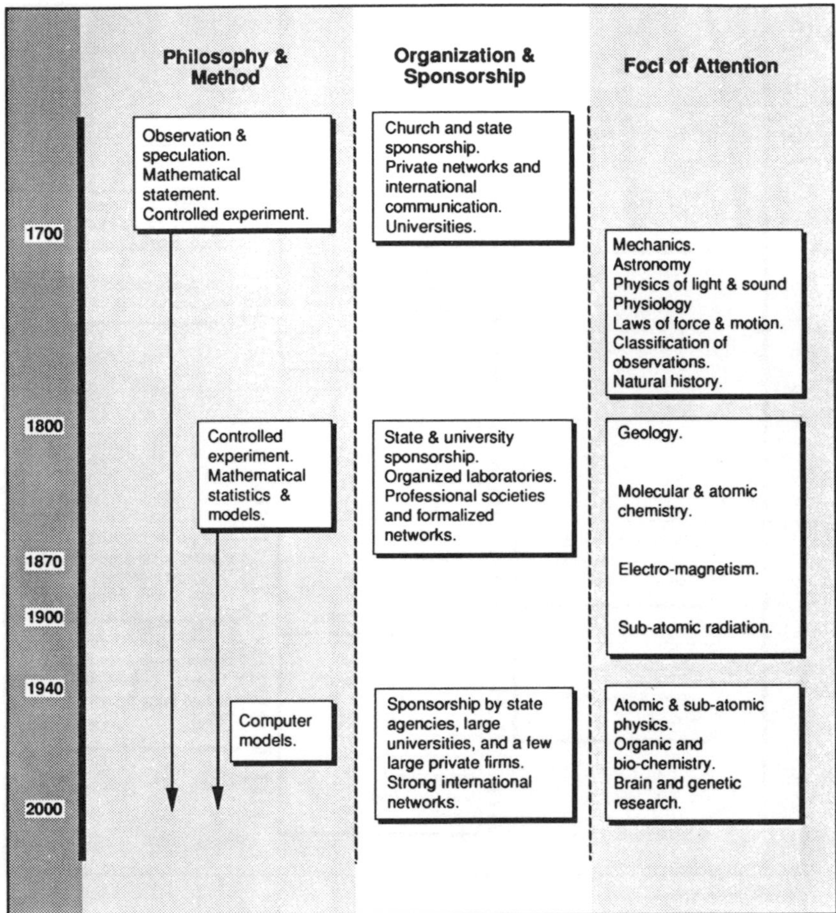

Figure 11.1a Scientific development: methods, organization and foci

ence, begins to feel the heat from its illuminations, first in the shifting focus, then more intensely as its method, its range and its immediacy make themselves felt.

Both science and technology are conceived of as functioning before the nineteenth century as bodies of activity pursued without strong formal organization, the work of individual scholars or craftsmen. If anything, the scientific community was the better defined and intercommunicating one, linked in places with ecclesiastical institutions (of which the universities were the most prominent) frequently still employing Latin in written communication; a secularised priesthood which, at least in the Protestant view,

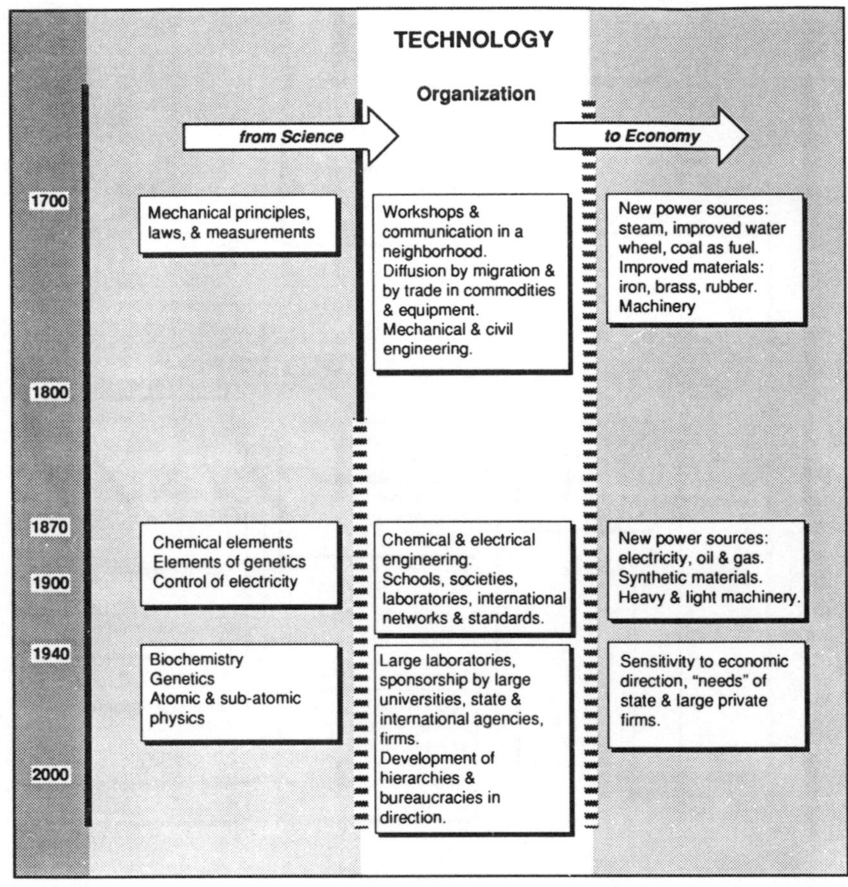

Figure 11.1b Technological development, 1700–2000

glorified God by demonstrating the reasonable and law-like character of His universe. It was only after a century, when the scientific attitude began to impinge on life and human creation itself, that difficulties arose. And as the power and wealth of the Church waned in northern Europe, the royal courts – and many princely ones – took its place as sponsors. The activity of scientific speculation became an acceptable one, too; what Veblen would have called 'honorific', for wealthy and intelligent members of the upper bourgeoisie. Some formal organizations in societies and colleges, with publication and interchange, began to create an international network. This loose organization grew wider and stronger as the scientific advances of the nineteenth and twentieth centuries proceeded.

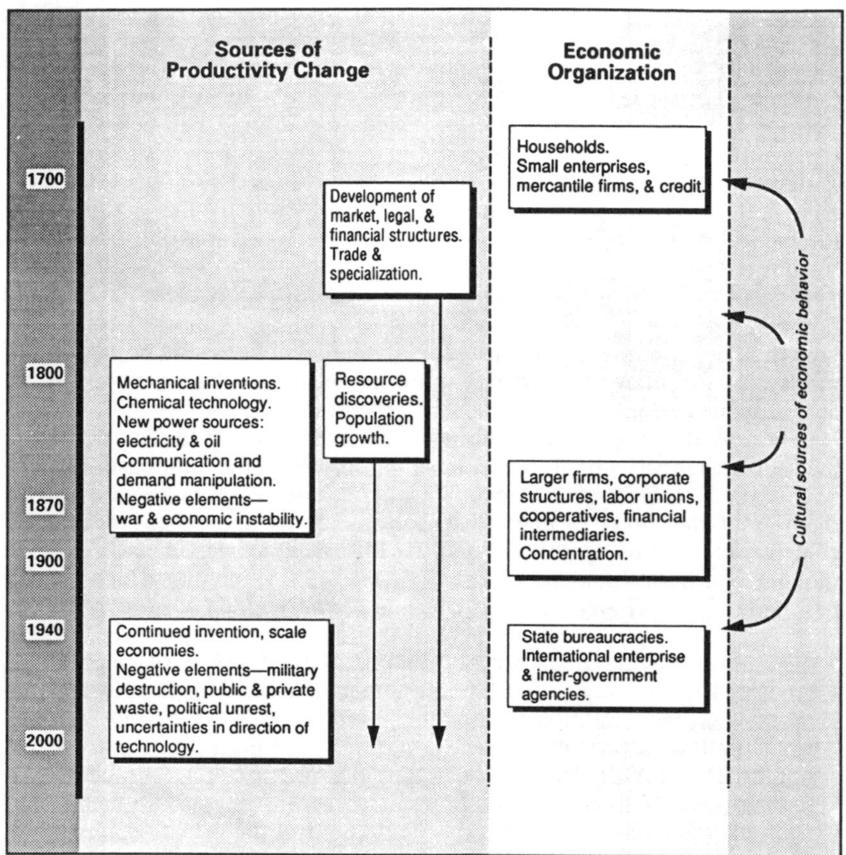

Figure 11.1c Economic development in Western capitalism, 1700–2000

Technological invention furnished many instruments and materials to a developing experimental science. At the same time, much of its own creativity, at least until the mid-nineteenth century, appeared in the random activities of isolated craftsmen and mechanics. Initially productivity depended on the sheer size of the inventive group, since the price of progress was innumerable trials and almost equally innumerable errors. As in all such primitive processes, where the inventors had no deep knowledge of what they were dealing with, the social model was that of a varied survival in the face of the destructive force of natural or market selection.

For this reason increasing artisan activity, particularly in a geographically limited and intensely intercommunicating region, raised the rate of technological change. It was, to be sure, essential that the activity be not all in one line, but distributed across the complementary materials and tasks of the production. Even though its main thrust was in mechanical inventions in the famous English 'Industrial Revolution', certain lucky chemical inventions, notably improvements in the making of iron and steel, permitted the replacement of wood as a structural material, and improvements in the design of coal burning furnaces were essential. Power was furnished by a more careful engineering of the water wheel, and by the extension of mechanical principles to a steam engine. This body of inventions, arrived at by different inventors within a narrow span of years, formed what Hughes might call a technological system. If one asked why it came in England between 1790 and 1820, the answer must be furnished from the economy: a growth in opportunity, in the presence of a large and growing population of mechanics and men of commerce. The model is the chain reaction of an atomic explosion in the presence of a 'critical mass'.

One cannot connect directly those early inventions to the learned arguments of scientists about planetary motion. Why the emphasis in both science and technology on mechanics, mechanical models, mechanistic theories of the effect of energy on solids, liquids and gases, mechanical contrivance? I must permit myself to borrow once again the pregnant thought of Charles Saunders Peirce (Peirce, 1891), beyond which no one has made much progress from its original formulation:[7]

> our minds having been formed under the influence of phenomena governed by the laws of mechanics, certain conceptions entering into those laws become implanted in our minds so that we readily guess at what the laws are. Without such a natural prompting, having to search blindfold for a law which would suit the phenomena, our chance of finding it would be as one to infinity. The further physical studies depart from phenomena which have directly influenced the growth of the mind, the less we can expect to find the laws which govern them 'simple', that is, composed of a few conceptions natural to our minds.

Were I truly a technologist, that concept, the concept of a capital structure – both of ideas and equipment – in science and technology, ever deepening as the deeper, more hidden levels of nature's secrets are touched, would have dominated my work and imagination; but it would require another lifetime and another education to follow it out.

THE RISE OF THE PROFESSIONAL ENGINEER

The social organisation of technological invention, unlike that of science, does indeed show a progressive change as the successes in the nineteenth century follow one upon another. 'Invention' as such is never wholly professionalised to where it can be taught and learned. It retains to this day its large random element; the distribution of the creativeness in a society, and its channelling among the branches of human activity even now, in our own specialised day, is only beginning to be subjected to scientific scrutiny.[8] Illumination even on the sources of illumination seems still to await its creative illumination. Here perhaps the scientific method may finally be caught up in itself and come to a terminal infinite regress.

Nevertheless, beginning even in the late eighteenth century, each of the successive lines of invention – mechanical, chemical, physical, biological or sub-atomic – develops a professional intermediary between science and industry in the role of the engineer. Ross Thomson's contribution to this volume marks indeed a phase intermediate between the craftsman and the engineer, especially in the machinery industry, where the mechanics in the producing firms formed an 'invention industry' themselves, intercommunicating and making developments that fed into one another.[9] But the application of principles of force and torsion, the known and measured qualities of materials in the design of a building, a monument, a bridge, a roadway, a mill or a canal had constituted the trade of the civil engineer from the time of the Pyramids, through the great achievements of the Romans, to the Renaissance and the English eighteenth century. Armies and dockyards had many such problems to solve, and men were trained in Napoleon's France at St Cyr and, in imitation, at West Point for such work. And each major body of technology after 1850 produced a counterpart: mechanical engineers, metallurgical and mining engineers, electrical engineers and, after 1940, the atomic, electronic and genetic subspecies. Engineering was attached to scientific discoveries or to the creative imaginations of investors as by a cell wall through which relevant principles, measurements and ideas could seep, as by osmosis, and be absorbed to be worked on in specific concrete applications. Those applications in turn derived largely from the investment and operating decisions, made in the economy by entrepreneurs, government agents and the men of affairs. A great semi-learned profession, intermediary between thought and action, formed a lower level of the priesthood of science. Engineers were indeed the parish priests, tending to the daily needs of a congregation on the basis of common sense, a few tricks learned as apprentices or at school, incorporating some of the ultimate skills, theories and illuminations teachable by the few to the many. Medicine is the most ancient and respectable of such arts, and the doctors to modern technology formed the strictest, the strongest and most respected element of every branch of an industrial civilisation.

TECHNOLOGY AND PRODUCTIVITY GROWTH

Productivity is the key quantity in the examination of the economic record of industrial societies (see Figure 11.1c). It is a concept borrowed from engineering and from accounting: the relation of the stream of output to the materials and forces consumed as inputs in the act of production.[10] It is the economist's equivalent to the engineer's concept of efficiency; measured, however, not in physical units, but in terms of value. But how can this ratio, on which a society's income and material welfare appears to hang, improve without some change in the technology of the transformation?

Logically, there are a number of possibilities, all involved in the definitions of input, output and social welfare. Economic historians have distinguished three that made their appearance before the speed-up of technological change and its closer linkage to the economy after 1800: (a) the growth of trade, wherein the purely commercial art of exchange distributes goods to where they are most highly valued; (b) changes in work organisation, in particular the specialisation of labour and equipment permitted by larger markets; (c) the so-called 'economies of scale', including those of 'scope' and continuous flow, that a large stream of production enjoys, where the sets of skills and fixed equipment in production can be continuously utilised.[11] To these well-known effects – the so-called 'Smithian' effects of the growth of the market – must be added a fourth, often neglected by economists working with the Ricardian model of diminishing returns: the geographical extension of the trading system, encompassing areas of improved resources: rich lands, dense forests, large and cheap bodies of minerals. This forms the most sensational and dramatic source of productivity growth in the nineteenth century. Accompanying capitalist expansion, too, some writers would put great importance on the development of the institutions of capitalism: assured property rights, courts and means of enforcing contracts and settling disputes, banks and the provision of flexible and responsive supplies of money and credit.[12]

In these non-technological, or pre-technological, historical processes of wealth creation, some technological change is both required and encouraged. Adam Smith notes how the subdivision of operations in a factory is conducive to mechanisation.[13] The employer simplifies the task of the inventor. The growth of sea trade produces many small improvements in ship design and materials. And, most notably, the expansion of agricultural settlement into new lands requires almost continuous innovation in field methods, implements, seeds and livestock; indeed, some of the improvement is taken care of by nature herself as she selects out the seeds, the genetic strains and the human stock which will survive.[14]

These purely economic sources of productivity growth in the nineteenth and early twentieth centuries are continuously renewed as a few major, startling inventions occur, especially in power generation, transport and communications. The railway first, then the telegraph and telephone, then the motor car allows massive reorganisations and relocation of productive operations, enabling a sensitively adjusting economic system to utilize them to maximum advantage. Yet there is no doubt about the importance of the pile-up of technology after 1850, as inventions, new machinery, new processes and new energy sources appear in every line of production. From 1870 onwards, the economic system is closely linked with its agents of technological change. Malfunctioning of markets, weak entrepreneurship, disruptive state activity and particularly shaky financial structures can slow down the rate of improvement, and disrupt the steady growth with crises, striking waste, civil strife, wars and prolonged depression. But the engine of growth still turns, and capacity, or the opportunities to create capacity, still grow: at least through the 1920s, and with a burst of revival between 1945 and 1970.

LARGER-SCALE ORGANISATION AND MELT-DOWN

The growth in scale, scope and geographical reach of the international market economy after 1870 produced some of the same effects that the earlier growing density of workshops, population and close communication had made evident. At the same time, the development of the link with science by way of engineering made innovation less risky. Technologists, as it were, came to know a lot better what they were about. Testing, experiment and analysis could replace the earlier methods of search. Countless failures were no longer needed for one success. A substitution of science for random chance, of brains for sheer energy and persistence, was making itself felt.

In such a situation, the natural reaction of entrepreneurs and their financial sponsors was to extend their reach into areas where the risk of failure could be easily borne. One observes, then, in all three diagrams, the growth of larger, more formally organised, non-market corporate systems in the economy, in government agencies and in the universities. In science and engineering, too, the twentieth century saw the growth in the scale and complexity of experimental equipment and laboratory installations. Sponsorship, through universities or directly by government agencies, or by large firms, came in large lumps. The three strands of the history, while preserving some separate identity, begin to fuse together. The twentieth century's two great wars had this effect, as science, technology and much of the economy was pressed into the service of state power. In the intervals of prosperous peace – the 1920s, the 1950s, 1960s and 1970s – along with high and effective activity by national governments, a growth in scale in economic organisations continued, and the functioning and growth of these two features of the organisational environment have continued unabated, along with great technological progress despite political slogans of 'privatisation' in the troubled 1970s and uncertain 1990s. The development was phrased with surprising pungency early on, in the farewell message of one of the most notable bureaucrat-warrior-managers of all time, General of the Armies (later President) D.D. Eisenhower, in a warning against a take-over of the whole national society by 'the military–industrial complex'. With this development on a world scale in the 1980s, all the creative activity formerly lodged separately in the intellectual structures of science and the professional organizations of engineering came to be drawn into the efficient service either of state power or the power of large-scale private economic organisations. One is tempted to conclude with the last chapter of *1066 and All That*, the famous English school-boy's parody on history:[15]

Chapter LXII
A BAD THING

America was thus clearly top nation and history came to a . . .

But that was written in 1931. Now, we know that at least Japan will be around, and Germany too, as well as – one hopes, in some form – the European Community.

No one would gainsay the great gains in knowledge and in wealth which a span of two centuries of almost continuous technological revolutions, in one category of production after another, has demonstrated. Still a fear, or the shadow of a fear, comes to all of us at times, like a bad dream in the middle of the night. If the creative impulses expressible in intellectual activity, in artisans' contrivance, in the enterprise of many competitive businessmen – these vents for energy and self-expression – are melted down together in a massive intellectual, governmental and financial bureaucracy, goal-directed in the service of state power and corporate wealth, what is left? What is left of the pure love of an activity, of the delight of exploration and discovery, the obsessive urge to achieve clarity, beauty or novelty of which earlier scientists and inventors were seized? And what becomes of the accidental, serendipitous breakthroughs which they hit upon? Perhaps vestiges of the earlier spirit – the rebelliousness, the love of adventure, of emotionality, irrational worship and unexplainable joy – linger on in the pockets of local human cultures beneath the blanket of a rational, power- and greed-driven materialistic, international or non-national 'high' culture. It is symptomatic that the resurgence of nationalism, tribalism, localism and the insistence of ethnic identity and family history – atavistic as these all seem to be, and so dangerous to a peaceful, efficient, smooth-running world – even these currents in the news seem to express the survival of the thirst for love, for contemplation, for generous activity, for knowledge, for purity and warm surroundings, as well as for excitement and the joys of struggle and of triumph. The problem with the emotions as part of our sense of self-worth is that, when society does not provide an outlet, they go sour and transmute into hatred, self-loathing and wild criminality. Fascism taught the world this lesson between the wars, even as the collapse of Leninist communism teaches it to us today. When knowledge, invention, economy, state policy and corporate wealth are all melded together, turning the hard and merciless face of the accountant, the banker, or the drill-sergeant towards humankind, history teaches us that there is hell to pay.

To arrive at the history of the underlying culture by way of a panoramic trip through only three human activities – science, technological change and economic development – may seem to be a perverse and roundabout route.[16] But scholars must persevere in their studies until they reach this point. The pressing need – the thing professors need to *do*, as Mr Dooley would have put it – is to define a humane, even a humanistic, value system, as wide as the world and as deep as the human heart. The effort need not limit itself to the cultural heirs of Homer, Plato and Aristotle, or Julius Caesar, as some vestigial classicists in the American educational establishment (nostalgic for the educational ideals of imperial Britain) would favour. Beyond science, beyond technology, beyond wealth and power: what balance in human nature, expressible in social life, will produce happiness? Voltaire, a philosopher friend tells me, observed that, 'to be happy, it is not necessary to be European'. But it *is* necessary to be *some*thing, something real, something limited and specific, even as one does not exclude from the range of one's sympathies the rest of humankind. That may not yield 'happiness', in some one of its Western versions, but at least it allows the full exercise of the whole range of human powers in whatever cultural milieu one lives.

A chapter on technology may seem a detailed, formalistic way to express intuitions that appear in the language of the underlying culture, where *human* values still linger and persist, yet perhaps such an exercise may be a way to break through to the human hearts of scientists, engineers and economists, proud of their *Wertlosigkeit* (worthlessness).

Notes

This chapter represents a formalization of comments made on a panel at the Conference on 'The Process of Technological Change', held at the New School for Social Research, New York City, in October 1989. I am indebted to Ross Thomson for patience and a little prodding, and to Heather Salome and the Mellon-West European Project at Yale University for contriving and funding, respectively, the preparation of the diagrams. The structure is an elaboration of my presentation of these topics in lectures at Yale and occasional seminars elsewhere.

1. Alfred Chandler's vision of the evolution of the large corporation and the national economic systems in which it has been encased is contained in three massively-researched volumes (Chandler, 1962, 1977, 1990). A review article on Chandler's work is in preparation for the *Journal of Economic History* in which his progression from the firm level, the national level and the comparative national (or 'Western world') level is observed, and the effort is made to restate the fundamental assumptions and dynamic of the Chandlerian vision in the language of economics and political economy. It is interesting, I would suggest, for any of Chandler's readers to attempt this and to discover how far he is able to do it for himself. This is called 'learning by book reviewing'.
2. Hughes (1983, 1989). The striking feature of Hughes's work is the dawning awareness it gives, as one reads about the development of power grids, of the analogy between a technology system or net, and the systems of social organization held together by the human electricity of communicated messages and sentiments. This analogy is made explicit by Hughes in an essay based on the earlier work: 'The Evolution of Large Systems', in Bijker, Hughes and Pinch (1987, pp. 51–82).
3. My own work, like that of Chairman Mao, is contained in two little red books, *Europe, America, and the Wider World* (Parker 1984, 1991), which contain 'Thoughts' on technology and its career in European industry and in American agriculture (see Vol. I, Chs 2, 3, pp. 68–73 and Chs 8, 9; Vol. II, Chs 9, 10). See also the re-statement in Parker, 'Technology, Resources, and Economic Change in the West' in Youngson (1972).

4. Landes (1983). Landes's great chapter on Technology in *The Cambridge Economic History of Europe* (Landes, 1965), was re-worked and amplified in his well-known book, *The Unbound Prometheus* (Landes, 1969), retaining the same emphasis on the different national social climates in which industrial entrepreneurs, and with them inventors and engineers, could flourish. Much as in a Marxian system, but against a less stark social background, purely technical, natural and intellectual influences and obstacles are subordinated to a sociology of enterprise. The economist's concerns with technological change are handled with interest and good sense in the various writings of Nathan Rosenberg. See especially Rosenberg (1976, Chs 4, 5, 6, and 1982, Chs 7, 8). The recent survey by Joel Mokyr (Mokyr, 1990) covers a broad span, both of time and topics, with much insight. See also the valuable contribution by the economist. Richard Nelson, to this volume.
5. Usher (1932). Usher's thought and values are most clearly articulated in the introductory chapters to his *History of Mechanical Inventions* (Usher, 1954), especially Ch. 3. Their influence on the treatment contained here, so far as my own limited self-consciousness is aware, is seen in the model of history as a set of 'threads', defined and separated by close internal linkages within each 'process history': i.e., sequences where the steps are a succession of generically similar human actions expressed in the revealed record of the history, and linked each to its predecessor in a more nearly pre-determined fashion than to its surrounding wider environment. So-called 'path-dependent' history suggests a similar view of historiography, but placed in a wider environment of economic and social processes and effects. See Arthur (1985) and David (1988). The technique is in effect Usher's engineer's approach to historiography, 'scientific', analytical and quantitative, as opposed to the intuitive, visionary, intellectual or poetic. On Usher's method, see 'A.P. Usher: An Appreciative Essay', in Parker (1984), Appendix, pp. 257–66.
6. The fact that intellectuals commonly speak of 'two cultures,' meaning science on the one hand and the fine (notably the literary) arts on the other is a confirmation of this bias on their part. There are, of course, myriad cultures or mini-cultures, specific to families, localities or professions in a society. But the notable omission from the idea of 'two cultures' is the culture of the economy: the world of practical affairs.
7. The insight is contained in an essay reprinted with others, filled with equally haunting or tantalizing suggestions, in a collection edited by P.P. Wiener, *Values in a Universe of Chance* (Peirce, 1958).
8. Scientific work on the genetic and physiological sources of thinking and creation has mushroomed in recent years beyond the ability of any mind, already trained in patterns of even a short time ago, to keep abreast, much less to gain the distance needed for a balanced assessment of its value.
9. See Thomson (1989). Reference must be made also to the current, as yet unpublished research of Michael Edelstein of Queens of the City University of New York, on the growth of the engineering profession.
10. See Parker (1990) and the references there cited for a review of the history of the concept and a critique. The article is reprinted as Parker (1991), Appendix B.
11. An interesting instance of this was given in an early article by Douglass North (North, 1968) where the sources of productivity increase in shipping in the eighteenth and early nineteenth centuries derived not only from the possibilities of more complex routing patterns with a large trade, but also from the elimination of piracy, which involved a naval effort of considerable size and expense. See also Parker (1984, Ch. 11).

12. North and Thomas (1973), Chs 1–3, 8.
13. Smith (1976), Book I, Chs 1, 2, 3.
14. Parker, 'Agriculture', in Davis, Easterlin and Parker (1972), Ch. 11. Also Parker (1991) Appendix A.
15. Sellar and Yeatman (1931).
16. A more direct approach is, of course, through idealistic philosophy in the Hegelian tradition or, more particularly, that of the Frankfurt School. I have made repeated efforts to penetrate this body of thought, but have been as often turned back at the gates by the vocabulary. My current effort is fixed on a book by Habermas (Habermas, 1968, 1973) in French translation, *La Technique et La Science Comme 'idéologie'*. Another alternative is offered by a religion like Calvinism which has an articulated theology.

References

Arthur, W.B. (1985) 'Competing Technologies and Lock-in by Historical Small Events', Center for Economic Policy Research, Publication No. 43, Stanford University (mimeo).

Bijker, W., Hughes, T. and Pinch, T. (1987) *The Social Construction of Technological Systems: New Directions in the Sociology and History of Technology* (Cambridge, Mass.: MIT Press).

Chandler, A.D. Jr (1962) *Strategy and Structure: Chapters in the History of the Industrial Enterprise* (Cambridge, Mass., and London: MIT Press)

—— (1977) *The Visible Hand: The Managerial Revolution in American Business* (Cambridge, Mass., and London: Belknap Press), p. 608.

—— (1990) *Scale and Scope: The Dynamics of Industrial Capitalism* (Cambridge, Mass., and London: Belknap Press).

David, P.A. (1988) 'The Future of Path Dependent Economies', Center for Economic Policy Research, Stanford University, August (mimeo).

Davis, L., Easterlin, R. and Parker W.N. (1972) *American Economic Growth: An Economist's History of the United States* (New York: Harper & Row).

Habermas, J. (1968/73) *La Technique et La Science Comme 'Idéologie'* (Paris: Gallimard, 1973), tr. with preface by Jean-René Admiral from the German *Technik and Wissenschaft als Ideologie* (Frankfurt-am-Main: Suhrkampverlag, 1968).

Hughes, T.P. (1983) *Networks of Power: Electrification in Western Society, 1880–1930* (Baltimore, Md.: The Johns Hopkins University Press).

—— (1989) *American Genesis: A Century of Invention and Technological Enthusiasm, 1870–1970* (New York: Viking).

Landes, D.S. (1965) 'Technological Change and Development in Western Europe, 1750–1914', in H.J. Habakkuk and M.M. Postan (eds), *The Cambridge Economic History of Europe* (Cambridge University Press, 1965), Vol. V, Part I, Ch. 5.

—— *The Unbound Prometheus: Technological Change and Economic Development in Western Europe from 1750 to the Present* (Cambridge University Press).

—— (1983) *Revolution in Time Clocks and the Making of the Modern World* (Cambridge, Mass.: Belknap Press of the Harvard University Press).

Mokyr, J. (1990) *The Lever of Riches: Technological Creativity and Economic Progress* (New York and Oxford: Oxford University Press).

North, D.C. (1968) 'Sources of Productivity Change in Ocean Shipping', *Journal of Political Economy*, vol. 76.

—— and Thomas, R.P. (1973) *The Rise of the Western World: A New Economic History* (New York and Cambridge: Cambridge University Press).

Parker, W.N. (1984) *Europe, America and the Wider World*, vol. 1 (N.Y.: Cambridge University Press).

—— (1990) 'Understanding Productivity: The Ways of Economics and of History', *Journal of Economic Behaviour and Organization*, 13.

—— (1991) *Europe, America and the Wider World*, vol. 2 (New York: Cambridge University Press).

Peirce, C.S. (1891/1958) 'The Architecture of Theories', *The Monist* (Jan. 1891) reprinted in P.P. Wiener, *Values in a Universe of Chance: Selected Writings of C.S. Peirce* (New York: Doubleday), p. 146.

Rosenberg, N. (1976) *Perspectives on Technology* (Cambridge University Press).

—— (1982) *Inside the Black Box: Technology and Economics* (Cambridge University Press).

Sellar, W.C. and Yeatman, R.J. (1931) *1066 and All That: A Memorable History of England* (New York: E.P. Dutton).

Smith, A. (1976) *An Inquiry into the Nature and Causes of the Wealth of Nations*, R.H. Campbell and A.S. Skinner (eds) (Oxford University Press/Clarendon Press), Vol. 1.

Thomson, R. (1989) *The Path to Mechanized Shoe Production in the United States* (Chapel Hill, NC, and London: The University of North Carolina Press).

Usher, A.P. (1932) 'The Application of the Quantitative Method in Economic History', *Journal of Political Economy*, Vol. 40.

—— (1954) *A History of Mechanical Inventions* (Cambridge, Mass.: Harvard University Press), rev. edn.

Youngson, A.J. (ed.) (1972) *Economic Development in the Long-Run* (London: Allen & Unwin).

12 Transformational Growth and Learning: Developing Craft Technology into Scientific Mass Production

Edward J. Nell

There is a famous detective story in which the clue is the dog that did not bark. An apparent break-in has resulted in a murder, but to the dog there was no intruder; the entry was nothing out of the ordinary, thus suggesting an inside job! The history of economic thought presents us with similar clues in the multiplier that was not there, in the work of Ricardo and Marx and, indeed, other nineteenth-century economists. Why, when they had developed all the analytical tools necessary, did they fail to point to the dynamic processes of the multiplier? The problem is particularly striking since the direct and indirect labour embodied in a good *is* the employment multiplier for that good, and the set of labour values for the economy as a whole is the matrix employment multiplier for the economy. Why did Ricardo and Marx not take the comparatively simple step of examining the dynamic process of employment adjusting through the multiplier?

The dog did not bark because nothing unusual occurred; perhaps Ricardo and Marx did not define a multiplier because no such processes took place in their time when investment or exports varied. (Of course, such variations had effects, but they may have been different or irregular, depending on unpredictable events, such as bankruptcies.) This may be an important clue to the nature of early capitalism and, even more significantly, to the way capitalism has developed. Dynamic processes depend in part on the flexibility of production, which in turn rests on the kind of technology in use. And technology, in turn, developed as a result of learning induced by the characteristic problems encountered in operating the initial production system.

An idealized contrast of early and later capitalism can be sketched: early capitalism consisted largely of family firms and family farms operating production technologies that depended on the presence and cooperation of skilled workers, working together. Such an economy tended to run at full capacity, unless seriously disrupted by business failures; product markets tended to clear through price adjustments. Employment remained fixed in the face of fluctuations in sales (short of the bankruptcy level); when output varied it was through changes in the productivity of labour. But this system created strong incentives to change the methods of production, in particular to increase the size of operations and to establish greater control over current costs, especially labour. Towards the end of the nineteenth century the methods of mass production were widely introduced, as we shall see, partly in response to pressures created by problems in the working of the earlier technology. Besides lowering costs, the new methods provided a desired degree of flexibility; but their successful adoption depended on the simultaneous emergence of adequate finance and a mass market, since these new methods required large outlays of capital. The change to the new methods can be called 'transformational growth' for, once adopted, these innovations in technology changed the way the system worked, replacing price with multiplier adjustments and full utilization with normal excess capacity.[1]

FIXED EMPLOYMENT TECHNOLOGY COMPARED TO MASS PRODUCTION

The change from craft technology to scientific mass production has largely been examined from the perspective of total cost reduction (Maddison, 1982). This is certainly a major factor, but the attention paid to it has perhaps led to the neglect of other dimensions. Indeed, economists have paid little attention to the actual characteristics of production technology. Output is normally considered to be a 'function' of various 'combinations' of the basic 'factors': land, labour and capital. The variations in the qualities and features of these are not considered, and neither is it explained exactly how they are 'combined'. The often-cited 'laws of returns' do not fit coherently together (Sraffa, 1926).[2] Everything is discussed at the highest imaginable level of abstraction and, in fact, the real object of the argument is to explain the distribution of income between rents, wages and profits on the basis of marginal productivity. The analysis of costs and their relation to prices is derivative. By contrast, the input–output approach tells us something about the technical relationships, since the various inputs for each unit output are clearly set forth, but there is still no consideration of how, exactly, these inputs are combined, or what varies with what.

Yet this is just what has to be considered if we are to explore the dimensions of flexibility. Craft production has often been praised for its greater flexibility, compared to mass production, because craftsmen could often adapt product design to the customer's specifications. Yet this is only one aspect of flexibility, and not the most important when the survival of the firm is in question. So let us turn to an aspect of technology that has largely been overlooked, namely the extent to which the process of production permits inputs or costs (and even output itself) to be varied so as to adapt to fluctuations in the state of demand.

Fluctuations in demand, of course, are endemic both in early and in developed capitalism. They are to be distinguished from permanent changes in demand, although it often may be difficult to tell which is which. Fluctuations may be quite temporary and local, temporary and global, long-term and local or global. There are seasonal fluctuations and variations reflecting local or temporary conditions on the one hand, and those of the general business cycle on the other. They may be foreseen, or unforeseen; if foreseen, only the direction or both their direction and magnitude may be correctly anticipated. But, even if fully and correctly anticipated, the realization may have dawned too late to do anything about it; or the duration of the fluctuation may be too short to be worthwhile adapting to. This does not depend only on the ability to anticipate the market; it also depends on the technology and organization of production, for it is production which has to be adapted.

Examples will help to indicate some of the ways in which businesses might try to adapt to variations in demand: if there is no refrigeration or method of storage, the whole current supply of fish and vegetables must be offered for sale, otherwise it will spoil. If demand has fallen, supply on the market cannot be reduced, so price will be forced down. Sometimes output cannot be varied; spring planting (and the weather) determines the harvest; if demand has dropped in the meantime, output cannot be changed, although grain can be stored, so supply could be changed, but at a cost. In a traditional blacksmith's shop the forge must be lit, and the apprentices on hand, whether much or little work is to be done. Energy and labour costs will be the same for quite a wide range of levels of daily output. Lighting the forge or, in steam-driven factories, building up a head of steam is time-consuming and labour-intensive. Thus, faced with fluctuations in demand, the methods of production may or may not permit the supply offered to the market, the output produced, employment and energy costs to be varied *pari passu*. To the extent that all or any of these cannot be varied, excess of shortage of supply will exert pressure on prices. But employment is the key; if it cannot be varied then the largest part of current costs will be fixed, and output can only be made variable by changing productivity.

More precisely, if employment cannot be varied, then it may not be worthwhile to vary output when demand falls; hence prices will be driven down. Yet if employment does not change, money wages will not be much affected so real wages will rise. Clearly, these characteristics of the technology can exercise significant influence.

However, the technological rigidities just noted do not necessarily translate in any straightforward way into corresponding economic inflexibilities, for there are at least two institutional forms that employment relations, based on craft technology, took in the past or can take nowadays (for example, in developing countries): first, there is the domestic system in which capitalists, often merchants, 'put out' work, paying piece rates to craftsmen working in their own homes and using their own tools and equipment. Then there is the factory system, in which workers are assembled in central buildings, owned and equipped by the capitalist, to work under direct supervision. The latter has many significant advantages. It permits training and close supervision, with the establishment of work norms and both product and labour standards, thus providing quality control. Labour skills in general will be raised towards the level of the best practice workers. Economies of scale will be realized, and machinery can be run off a central power source, such as steam or water. And it will eliminate the sometimes costly and bothersome travel to and fro for the delivery of materials and collection of finished work.

In each case the technology permits little adaptation to variations in demand; current costs are the major portion of costs, and the greater part of current costs are real and fixed. But the domestic system puts this burden on the craftsmen and their households, whereas the factory system obliges the capitalist to assume it. As we shall see, the early factory operated as a (largely) fixed employment system, with little ability to run on short time or at less than normal capacity. But in the domestic system, when there was less work to be put out, the craftsmen, having to support their families and establishments, could be expected to bid for the available work, driving piece rates down and putting the least efficient workers out of business altogether. This system therefore provided early capitalism with some flexibility in response to variations in demand, thereby giving greater protection to the class of financial rentiers whose fixed interest income depended ultimately on profit receipts.[3]

Two Systems of Technology

To see the contrasts sharply, compare 'pure' stylized cases. In a craft or fixed employment system workers control the pace of work, and their skills, using general purpose tools, hone the usually customized product and define the details. Under mass production, the machinery determines the pace of work, and the precision is built into the equipment, which is designed specifically for the product. As an example of the first kind of system we will take an idealized economy of early factories and artisan shops, in which small family firms and farms practise traditional crafts, drawing on the traditional sources of energy, such as wind and water, animal and human effort, but also steam power. The system is capitalist, in that a (more or less uniform) rate of profit prevails and governs investment, but although workers will be gathered into factories, scientific mass production has not begun. Craft methods are practised by teams of workers following long traditions. Craft work requires the presence of the entire work team, full time, for there to be any output at all. Start-up and shut-down costs are typically large, as are storage costs. The little technical progress that takes place is unsystematic, and the few economies of scale that exist are offset by increasing risk and diseconomies. (This picture deliberately overemphasizes the fixity of employment, and neglects the flexibility provided by the 'domestic system' discussed earlier. That system was a relic of pre-capitalist industry, retained *in spite of its disadvantages* because of the flexibility it provided (Mantoux, 1928). To understand the economic relationships of the new system of early factories, and the characteristic problems facing this economy, we must set aside this mixture and consider the pure form.)

In such an economy, in the long run prices must cover costs (including normal profits) and so will reflect distribution, but in the short run employment will be relatively stable since work teams cannot easily be broken up, with the result that a high proportion of current costs will be fixed.[4] Needing to cover these costs and lacking technology for storage and preservation, when demand weakens, goods will have to be sold for whatever they will bring, so that prices will adjust to the requirements of market-clearing. Employment and output will tend to be stable, with prices flexible. (Note that 'putting out' effectively turns a craftsman labourer into a small business unit, so that piece wages behave like prices. The worker households have large fixed real costs and so must, in the short run, compete for whatever work there is to be had, as long as the pay for such work is more than the variable costs associated with it. Hence as piece rates vary with demand, and therefore with prices, real earnings rates will tend to remain steady, but less efficient – or less well-placed – workers will find their employment varying with demand.)

Artisan production may have a well-developed division of labour, and it may also be highly mechanized (Sabel and Zeitlin, 1985). (Adam Smith's pin factory has the first, and Marshall's examples of the printing and watch-making trades (1970) exhibit the second as well.[5] Prior to mass production the use of machinery replaces the workers' energy; but the early machine system drives essentially the same tools, although on a much larger scale.[6] Acceleration and deceleration will be slow and difficult; transmission of power will be handled by belts and pulleys in manner both cumbersome and inflexible. Such use of machinery does not affect the system's characteristic mode of operation, for neither division of labour nor mechanization will have proceeded to the point where the pace and quality of production is controlled by the machinery, as in the science-based assembly line (Howell, 1991).

In the early factory/artisan economy the work team must be kept together; everyone works or no one works. In the extreme case, lay-offs will simply not be possible; the factory or shop must either operate or shut down altogether. If machinery is used it must likewise all be used.[7] Work is skilled and workers have to coordinate their efforts; workers themselves largely set the pace of work, so that productivity depends heavily on morale. Many technical aspects of production are not (and given that craft production is customized, cannot easily be) written down, but exist in the minds and accumulated experience of the foremen and senior skilled workers (Argyrous, 1991, Broehl, 1959). Technological improvements thus accumulate as specific skills and specialized knowledge on the part of senior workers and foremen.

A work crew that functions well together will be highly important. Moreover, most production will be discontinuous 'batch' production; a batch, once begun, must be completed. But a good work crew should not be broken up, so new batches must be started even if demand is weak. Even when early forms of continuous production have been instituted, however (especially with the use of steam power), start-up and shut-down costs will be significant, so that the firm cannot go on half-time or close down for part of the week.

The inability to lay off labour has important consequences in several areas. Firms accumulate financial reserves during booms and good times, which they use to tide them over hard times. Reserves are designed to smooth over the cycle; they are not really available for investment. In planning production, firms will be reluctant to contract out for parts or repairs; they would rather use their own labour and facilities in slack times to do such work. (And their own foremen can supervise, which may be important when designs are not fully drawn.) As a result economies of specialization will be lost.

Finally, products are non-standard; they are made to order, which means working procedures will frequently have to be reorganized. Moreover, few products can be produced to stock; storage facilities are both poor and expensive, so that losses in terms of wastage and decay will be heavy.

The system of mass production differs in every one of these respects; products are standardized, and storage facilities climate- and pest-controlled. The pace of work is governed by the speed of the assembly line or other machinery; jobs are broken down into their simplest components, reducing the need for skill. Skills remain important but, ideally, no worker need have more than one basic skill, to be exercised repetitively, in conditions where precision equipment eliminates much of the need for judgement or timing. Tasks are regularly simplified through time and motion studies, conducted by trained engineers. Work is continually reorganized, resulting in a persistent, though variable, tendency for output per worker to rise. R&D is carried out by a staff of professional scientists, who remain in contact with universities. Very little technical knowledge will remain specific to foremen or workers, as in the craft system, although aspects of technology may still remain firm-specific. The labour force has no need to interact, so morale counts for little, and workers are dispensable and interchangeable. Finally, start-up and shut-down costs are minimal; power is provided by petroleum-based fuels and electricity. These differences in technology, energy and labour requirements make possible a very different form of market organization: the corporate industrial system of mass production. (Chandler, 1990). (In practice, of course, elements of the early factory/craft system will remain, particularly in the areas of agriculture and primary production. Even more surprising, though, is the fact that many industries will shift to mass production only under the hot-house pressure of war-time mobilization. Aircraft and shipbuilding used craft methods until 1939; aircraft production ranked 41st in US industries in 1940 and employed 64 000 workers; by 1945 US aircraft manufacture was the largest industry in the world, and employed 2.2 million workers: see Argyrous, 1991.)

Under the corporate industry form production is carried out by corporations, organized as large bureaucracies, the ownership of which is decided in the financial markets. Technical progress is regular; economies of scale are widespread. New products and new processes are frequent and innovation is one aspect of competition. The size of an operating unit in the artisan economy will be limited by difficulties of coordination, cost of transport for distant markets, storage costs and increasing risks; in many cases, craft technologies will dictate a 'natural' size for the plant, and family firms will tend to operate a single plant whose size cannot exceed the requirements of an optimal work team. In the case of mass production, however, a larger plant confers significant economies of scale, which certainly must be balanced against increasing risks and distribution costs. But plant size does not determine output capacity, since throughput can be speeded up by the re-design and reorganization of work. Still less does plant size determine firm size. (Notice that there will also be economies of scale in storage, such as canning, drying, refrigeration, grain storage and so on.)

Under mass production growth is inherent in the system, and investment is carried out as far as possible by existing firms, who will not leave it to newcomers. New entrants would threaten the arrangements in existing markets since new equipment will normally be superior to old, providing newcomers with a competitive edge. Hence firms will do their own saving, in the form of retained earnings. Price guidelines will be set so as to earn just enough on normal operations to finance the expected required investment. Actual prices will be held close to the guidelines, fluctuations in demand will be met by adjusting production and employment.

This means, however, that any given change in demand, by causing a corresponding change in output and employment, creates a secondary change in demand in the same direction, but smaller in magnitude. This secondary change, in turn, sets up a tertiary, and so on, until the effect is no longer noticeable. The initial variation is therefore *multiplied*. But this multiplier is rooted in the technology and cost structure of the system, for that is what chiefly determines the proportion of revenue that is passed along; it has nothing to do with the 'psychological propensity to consume'.[8] Gross profits are 'withdrawn' at each stage, since they enter business financial accounts, and households may also hold back a portion of income. 'Saving', however, meaning a positive decision to accumulate assets (whether by businesses or households), is not the same thing as 'withdrawal', and it is the latter that is relevant; the crucial question is what proportion of revenue continues in circulation.

Let us draw the contrasts now: the mass production system is able to maintain productivity while varying output. The artisan system, by contrast, operates with fixed capacity utilization and variable productivity, while mass production has variable capacity utilization and fixed productivity at the margin. Hence an artisan process will have low variable and high fixed costs, but the fixed costs will in large part be current costs, chiefly labour, set in real terms, rather than capital costs fixed in monetary terms. The industrial economy, on the other hand, will have much higher variable costs, and its fixed costs will be capital costs set in monetary terms.[9] In the artisan economy, therefore, changes in demand may lead to changes in the intensity with which workers work, but not to changes in employment. If demand increases, then, output per capitum may be increased, although output per unit of effort may actually decline; indeed, it normally will decline after a point, on the assumption that, with given plant, returns to additional effort diminish. Conversely, when demand declines, output per capitum will decline, although output per unit of effort could remain constant or increase.[10]

Under mass production a variation in demand, due to fluctuations (for example, in investment), will be amplified: the changes in investment demand will result in changes in employment, so that consumption will change in the same direction.[11] By contrast, in the early factory/artisan economy a change in demand will result in price changes. Fundamentally such an economy has only two levels of operation: all out or zero. To cut back output without closing down will be difficult; work must be done more slowly but for the most part, the entire crew will still be working. Given time, work can be reorganized, so some variation in employment is possible, but given an organization of work the only way to adjust output is to vary productivity at the margin. Since this will seldom be profitable, a fall in demand will result in excess supply, pushing down prices. This needs careful explanation: in the early factory/craft economy, with employment largely fixed, worker consumption will be governed by real wages. Money wages will be set at the time of employment and, since employment tends to be relatively constant, money wages will not vary much either. Prices, on the other hand, will reflect the need to earn as much over variable costs as demand and the competition permit; hence, when demand is below normal, competitive price cutting will take place and, when it is above normal, prices will rise. Prices will therefore vary flexibly with demand.[12] Hence real wages will move inversely to variations in demand. If the chief cause of variations in overall demand is investment, reflecting 'the general state of business confidence', then the induced change in consumption will tend to be in the opposite direction to the initiating change in investment.[13] Hence a decline in investment demand, so long as it is not too large (leading to bankruptcies), will lead to a rise in consumption, which is just the reverse of what happens in mass production. The system has a built-in stabilizing mechanism, so long as the fluctuations are not too large.

Prices and Demand

Long-run or normal prices in a capitalist economy can be expressed by 'normal cost' equations: the price of one unit equals its wage cost plus the input costs plus the profit. Breaking this down shows the coefficients of the various inputs into each process, each multiplied by the amounts employed and by its price, aggregated to get total input cost. This figure will be multiplied by the gross rate of profit – the net rate plus the rate of depreciation – and then combined with the labour inputs times the wage rate. This results in the familiar 'Sraffian' or classical matrix expression for prices and the rate of profit, given the wage.[14]

This equation shows how the basic planning or benchmark prices, set at the time when investment projects are decided, interact with each other and the rate of profits, but it does not tell us how firms settle on the prices they will actually charge in changing conditions, and neither do the coefficients tell us how the system of production operates. To arrive at an account of how firms vary their prices in actual conditions it is necessary to know more about the technology. In particular, we have considered two idealized systems: an 'artisan' economy and an 'industrial' economy. Each can be considered a complete economy, capitalist, described by the classical equations and capable of growth. But the behaviour of costs and prices and the pattern of growth will be different (Leijonhufvud, 1986). Moreover, the basic 'normal' price equation will have to be interpreted differently.

In the craft economy long-run prices are independent of demand (apart from indirect influence in cases of joint production and land, which involve switching techniques); they depend on technological coefficients and on the level of real wages, and, of course, on the relative degree of competition in the various sectors. In strongly competitive conditions, capital will shift in response to differentials in profit rates, tending to establish a uniform rate. Long-run prices are set at the same time investments are planned, and function thereafter as benchmarks or guidelines. But short-run prices under a fixed employment system reflect current market conditions: that is, the balance of current supply and demand. They are highly volatile precisely because costs, and to a lesser extent output, are inflexible.

In the mass production economy, however, while prices are also set in the course of planning investment and are independent of the level of current demand, they are not independent of the growth of demand. Indeed, the rate of growth of demand is fundamental in the sense that prices have to be set to cover the costs of the investment necessary to construct the capacity to service the new demand, by the firms presently in any given industry.[15] 'Entry' is chiefly a phenomenon in new industries, and will be associated with variations or improvements of a new product. This will lead to price equations that are almost identical to those describing the craft system. Interestingly, however, the balance of current supply and demand has little or no impact on prices; market prices tend to stay near the benchmark levels, and variations in demand chiefly affect output and employment. This is due partly to the fact that output and employment are variable, but it also reflects the shift in the nature of competition, from a concern with prices to a race for new technology (Eichner, 1976; Wood, 1978; Nell, 1991, Ch. 17).

A Digression on Methods of Analysis

The so-called 'long-run method', common to the classics and the early neoclassicals, is the appropriate analytical approach for studying the changing configurations of economic variables in the craft economy. The classical procedure was to determine the prices and rate of profit, on the basis of a historically given real wage; then to establish the level of capacity that corresponded to the level of demand. The result would be the system's 'fully adjusted' position, which can then be compared to similar positions with different configurations of the basic parameters (Garegnani, in Eatwell and Milgate, 1983). (Of course, a neoclassical analysis would determine the prices and outputs together, rather than separately, and the real wage would be determined by supply and demand. But the neoclassical analysis, say, of Wicksell, was firmly based on the long-run method.)

This approach, however, is not well-suited to the study of the mass production economy. First, there is no reason to restrict output to a particular level. Since it is easily variable, it may settle on a path, a particular pattern of motion, rather than a level, even in the short run. The analysis would therefore have to be explicitly dynamic. Second, the final resting point may itself depend on the pattern of movement towards rest: the equilibrium may be 'path-dependent'. Certain paths, for example, may induce more or different technical progress, or may lead to more pressure on fixed resources, than others (Robinson, 1980, Ch. 7). Third, since technical progress, in regard both to products and processes, is a regular and endogenous feature of the system, equilibrium configurations are bound to be very short-lived, if they exist at all.[16] It is more appropriate to identify the trends, noting, however, that technical progress need not be uniform across sectors, so relative costs may be changing. Appreciation of these points undoubtedly contributed to movement, by Keynesians and neoclassicals alike, away from the long-run method and in favour of 'Temporary Equilibrium', often treating growth paths as successions of temporary equilibria.

Nevertheless, there will still be comparatively stable benchmark prices which will have to be determined, and which will reflect the planned prices at the time of the investment decisions that created the presently existing capital structure. These prices will be expressed by the same classical equations used in the long-run approach, adjusted for 'vintages', but the interpretation will be different (which will explain the comparative stability in the face of continuous change). The 'long-run' classical equations are defined on the basis of the 'best-practice' coefficients, since the resulting values are the ones towards which competitive pressure are forcing the system. But this overestimates the actual size of the surplus, and therefore the rate of profit. It would also imply continuous and often large price changes. Under mass production technical change is continuous, and firms are constantly rebuilding or renovating plant and equipment; the purpose of the equations is to exhibit the benchmark values and determine the size, composition and distribution of the surplus. Competitive pressures are not able to enforce any tendencies since, even as they begin to act in a particular direction, new technical innovations will open up new opportunities. Moreover, the new improvements will be in the renovated or new equipment of existing producers; it will not be in the hands of new firms, eager to put older firms out of business. Under mass production the growth of the market is supplied by the expansion of the firms already serving the market, so they will set prices on the basis of their *average*, rather than their best-practice, coefficients of production. The average, of course, changes much more slowly, and actual prices (as shown in inter-industry studies) are quite stable over time and remain close to direct plus indirect labour costs (Ochoa, 1984).

Summary

Real world economies will always be a mixture; very likely no economy was ever purely a fixed employment or purely a mass production type. But the two systems differ profoundly in their mode of operation, a point which shows up most clearly in the way markets adjust to variations in demand. In the mass production economy, adjustment takes place through the multiplier, but in the craft economy the adjustment process works through prices and distribution – akin to the 'Cambridge' theory of distribution, but also with affinities to Marshall – and the difference rests on the different nature of costs in the two systems (Nell, 1991, Ch. 16).

TWO PATTERNS OF GROWTH

No doubt artisan technology could be operated in stationary conditions but, if the economy is fully capitalist, profits will have to be invested since the driving motivation will be to accumulate capital. But the characteristic pattern of accumulation will be through the lending of household savings to new firms, which will set up shops that replicate existing ones, but serve new customers. Firms expand to an optimal size and operate at that level thereafter. However, the system's prices do not depend on growth. (In this sense an artisan economy behaves as if it were stationary.) By contrast, in an industrial economy firms retain profits and invest in expanding and improving their own facilities and set their prices to support the requirements of this growth. The reason for these differences in the pattern of accumulation lies in the different relationship of technology to competition in the two cases, and hence to differences in the nature of firms. In the first case firms are stationary, but in the second they grow regularly.

In the artisan economy success in competition comes through the development of the skills and morale of workers. The successful firm has the better product, more reliable delivery times and quicker production times (with unit costs therefore lower), and so on, all of which depend on worker skills and their ability to function together as a team. Such characteristics are personal and intangible; improving them does not depend on rebuilding factories or re-equipping shops. Very often crucial details of the technology are never written down; they reside in the minds of the foremen and senior craftsmen. But this means the technology is unreliable; sickness or disaffection of key workers could undermine the whole year's effort.

Learning and innovation are therefore confined to work teams. An innovation may arise within a group; but keeping it secret will provide a competitive advantage. Diffusion will usually only take place as workers move from one area or company to another, or as foremen leave to set up their own companies. Diffusion and the passing along of information was difficult not only because no one tried to write down such improvements, but also because, prior to the rise of professional engineering, the standards of industrial drawing and technical writing were low and irregular, so that the competence to do so was often lacking.

In the industrial economy competitive success likewise depends on cost-cutting and improved product design, but the difference is that these are objectively grounded in the production process, rather than based on intangible personal characteristics. For the most part the technology will be written down; indeed, it will be based on science, and on a professional engineering culture centred both on universities and on private research laboratories. It is therefore reliable, but dependent on the way the technology is grounded in institutions and embodied in equipment. 'Embodied' technological innovations require retooling or rebuilding plant, and this necessitates investment. Plants have to be shut down and renovated, or scrapped and rebuilt. Even 'disembodied' technical change, however, requires redesigning the work flow and the organizational chart.

Regular technical change requires professional engineering; with the rise of mass production came the rise of professional standards and schools. Degrees in mechanical, electrical and chemical engineering arose; mechanical drawing developed; time and motion studies, flow and organizational charts were initiated; all in the effort to understand the process of innovation and bring it under control.

An improvement provides a competitive advantage and must therefore be matched, so there will be a need and an incentive for the economy to contain a sector which specializes in supplying the means of production, and which is large enough to meet the demand for rebuilding entire industries. Investment will be more or less continuous, and productivity will regularly rise, although not necessarily in step with demand.[17]

The saving–investment process differs markedly between the two systems. When the artisan economy expands, household savings will be loaned to set up new firms. The system of small establishments will replicate itself; growth will not normally be undertaken by adding to the capacity of existing firms, for there are few economies of scale in traditional crafts, while adding to size increases risk. (There is a problem, however, in the traditional account of a 'perfect capital market'.)[18] Industrial systems, on the other hand, do provide economies of scale, both in the design of equipment and in the organization of work. Even more important, however, technological competition between suppliers of capital goods means that new equipment is likely to be better or cheaper than old.[19] New investors will have an edge; existing firms cannot afford to remain satisfied with their present scale of operations, leaving growth to new entrants, for new firms will be able to undercut them in their own markets.

However, existing firms do not necessarily have to scrap and rebuild every time there is a significant innovation. This would be wasteful, both socially and privately. Instead they can adopt the innovation in building new capacity to meet growing demand, carefully building just enough – at an appropriate price – to prevent newcomers from entering. The industry will then consist of a number of firms, each having both new and old plants, rather than of older firms with outdated plants and newer ones with superior equipment.

To achieve this, however, growth must take a different path from the artisan economy. To avoid competing for household savings, firms will retain their earnings and invest them directly. So long as their investments are judged to be wise, the value of their equity will rise in proportion, which means the shareholders desiring funds can obtain them by selling off an appropriate part of their holdings at the higher price (Nell, 1991, Ch. 21).

Industrial growth thus differs fundamentally from growth in an artisan economy. Competition requires regular investment, financed by retained earnings, with a consequent rising price of equity. Labour becomes a variable cost, and output and employment vary together in line with sales, while productivity, fixed by technology, stays constant. Capacity capable of meeting the maximum likely demand can be installed, thereby ensuring that there will be no room for newcomers, without any risk of having to meet the labour cost of that capacity when it is not in full use.

In the artisan world, growth simply replicates existing stationary relationships. But growth is built into the working of the industrial system; for example, it must occur for potential profits to be realized, but (of even more relevance here) it is part of the competitive process. If markets currently clear, but there is an imbalance between rates of growth of supply and demand, they will fall into disequilibrium in the future. If they do not clear now, but their rates of growth are in balance, they will eventually even out. Prices, moreover, are significant in relation to rates of growth: at low prices new customers can adopt a new good, so demand can expand; but low prices mean low profit margins, so little finance for construction of new capacity. This suggests that a price might be found that would just balance the rates of growth of supply and demand. Generalizing this makes it possible to work out a pattern of price determination for a corporate industrial system, showing the relation of pricing to growth (Eichner, 1976; Wood, 1976; Nell, 1991, Ch. 17).

Mass Consumption

The corollary of mass production is mass consumption and, just as the new techniques of mass production require changes in jobs and business organization, mass consumption changes both the nature of household organization and the work done in the household. Mass consumption, first of all, entails standardization on the demand side. Besides creating a mass market for uniform consumer goods, repairs and spare parts will likewise have to be standardized; hence repairmen will require standardized tools and appropriate training. Delivery, repair and servicing systems will also have to become standardized and cover the entire market. Small-time individualized craftsmen and handymen will gradually give way to authorized dealers providing standard service, along with 'chain store' repair shops.

One consequence is an increased dependency of households on industry. Mass produced goods tend to be delivered in finished form; households do not have to put them together. The need for standardized spare parts means that repairs will be done by the servicing industry, so households skills and independence will be eroded. This increases the time available to households, and no doubt makes sense in terms of the division of labour, but by increasing interdependence it also increases dependency.

Of course, mass consumption also entails mass distribution, the development of coordinated transportation, storage and marketing facilities across the entire nation; a huge task requiring large, professionally managed organizations (Chandler, 1977).

With the advent of mass consumption all households will be consuming standardized products, whereas previously many goods were made to order, to individual specifications. But how did households all learn to *want* the same goods? Did they not have to give up at least part of their individuality? Once mass consumption has developed, individuality can no longer be expressed in the goods the household consumes or uses, or in goods which are consumed in complementary patterns with them, since these will also, for the most part, have to adopt a uniform character.

While individuality may be sacrificed (to some extent), however, mass production/mass consumption may even enhance social differentiation, for mass products can easily be made into universally recognized status symbols: in the USA, Fords and Chevvies for the workers, Buicks and Mercuries for the middle/upper middle class, with Cadillacs and Lincolns reserved for the upper crust.

This may give us a clue to what made households so readily give up the individuality of their consumption patterns, and all come to want the same things. Part of the answer no doubt lay in cost: mass-produced goods are cheaper, and if the distribution networks are adequate they may be at least equally convenient. But, if cheapness were the whole answer, we would have to ask at what price would people give up their individuality? How could business know the price at which *everyone* would, all more or less at the same time? Creating a mass market would be a matter of being able to mass produce at the right price, but how could business know this in advance?

In fact the development of mass markets is not just – or even primarily – a matter of price. Even more important is the growth of income and the development of new products and processes, and their impact on households. Mass consumption has always been encouraged through competition, but it is a special kind of competition, engendered by business, through advertising and sales campaigns; it becomes a competition between households for status, as displayed in consumption. Veblen noticed this early on, but considered it a matter of conspicuous consumption, although that was only part of it. The competition for status had a deeper foundation in changing employment opportunities. The emergence of mass production created new categories of employment, a middle class of engineers, professionals and white-collar workers, in between the traditional classes of owners and workers. (Indeed, in time the upper ranks of the middle class came, in part, to displace the owners as the dominant force in the economy.[20]) These new ranks had to be recruited out of the urban and rural labour force, and this fact in itself set up a race to rise in status. The method was self-improvement; households had to invest in learning new skills, in education and vocational training and in moving to take advantage of new opportunities, all of which meant advancing their skills, largely through education and (as a corollary) displaying their achievements through their consumption.

By contrast, in the early factory/craft economy, households (centred on a breadwinner) normally functioned in an unchanging manner, reproducing themselves by producing a new generation of children who would marry and set up similar households. 'Preferences' for the goods which support normal household activities would therefore be stable; conventional consumer demand theory may provide a reasonable account of this sort of household expenditure, and also aid in the study of household responses to price changes (although it does not provide an easy way to study the introduction of new products, a major phenomenon even in the strata of society that did not enter into the race for self-improvement).

In the mass production economy, however, for a significant class of households – those attempting to rise in the world – preferences will be changing, and in a regular way. The expenditure of these households will be governed by their plan to rise in status, and their choices will be made accordingly (Nell, 1991, Ch. 17). In the professional and managerial classes, competition for promotions and status will compel even households contented with their current station in life to enter the race, simply to defend their present positions. The result will be a spread of demand for those goods and services that help to achieve or come to indicate the desired status. Mass markets tend to develop first among the intermediate strata, spreading only later. The study of demand, therefore, requires a dynamic approach; the dominant theme will be *growth* in demand: both relative growth for particular goods, and growth of expenditures in the aggregate.[21]

THE PRESSURES FOR CHANGE

An early factory/craft economy, in our idealized version, has a pattern of stabilizing adjustments which depend on market-clearing prices. From the point of view of the system this may be good, but from the point of view of individual producers these adjustments have some undesirable properties. For example, when demand falls off, production will be run more slowly but, exceptions aside, the full labour force will still have to be on hand. From the individual owner's point of view this is an unfortunate expense, which necessitates injurious price competition. In the absence of a domestic 'putting-out' system, the burden of adjustment will fall heavily on profits: when demand falls and prices are hit, if employment cannot be changed and given a mark-up of 50 per cent, a 5 per cent drop in demand means a 10 per cent fall in profits. If two-thirds of the capital were external, a 17 per cent falling-off in demand could bring bankruptcy. Hence leverage must remain restricted under these conditions. Conversely, when demand is strong, the rate at which production can take place will depend on the morale of the labour force, and its willingness to put in extra effort. The capitalist does not have full or satisfactory control over the pace of work, or the level of costs.

What the capitalist needs, first, is greater control over the process of production, especially over the productivity of labour and the pace of work. Time and motion studies were developed to provide just this (Barnes, 1956). Such methods provided greater control over labour but, since they required professional management, at the price of sacrificing family control over the firm itself. Second, firms wanted to be able to expand and develop with their markets, not just to reach an optimal size and stagnate, and they wanted greater control over output (to be able to vary production with sales). They needed reserve capacity, and to be able to shut down temporarily. Third, and correlatively, firms needed to be able to vary costs when sales are varying, which requires being able to lay off and rehire labour easily which, in turn, depends on being able to schedule and re-schedule production. To do this, start-up and shut-down costs must be minimal. Fourth, firms need to be able to store output without spoilage. If, when sales fall, output can be cut back and, along with output, costs can also be cut, and at the same time unsold inventory can be stored without significant loss or other costs, a great deal of pressure for potentially ruinous price-cutting will be lifted. Part of the problem thus can be reduced to a technical question, which is how can production be run at less than full blast without all workers having to be present? Alternatively, how can production be started up and shut down, easily and costlessly, so that a drop in demand can be met by running short-time?

The normal working of the system throws these questions up; the answers will help businesses to compete more effectively. Inflexible output and costs, resulting in overproduction and cut-throat price competition, is potentially ruinous. The system itself thus creates the pressures which lead to the technological developments that make labour a cost that varies with output, which in turn varies with demand. The result is an improvement in the flexibility of the firm's response to changing market conditions.

Business will attempt systematically to gain greater control over the production process, substituting mechanical power for labour power, and mechanical or electronic control for human skill, as far as possible. In general the larger the scale on which operations take place, the better the prospects for doing this, which is an aspect of economies of scale: 'the division of labour is limited by the extent of the market'.

The chief method by which business can achieve these goals is to speed up throughput, and make it continuous (Chandler, 1977). Processes of continuous throughput, replacing or modifying batch production, provide a steady and adjustable flow which can be matched to the level the market requires. To make this possible, it is necessary to shift from earlier and more primitive energy sources – animal power, human power, water and steam – to electrical power or petroleum, especially the internal combustion engine. (The earlier energy sources all face substantial cost problems: water power is weather-dependent, often seasonal or unreliable; water and steam equipment requires substantial maintenance even when not in use; animal and human power require feeding and support even when not in use; and steam and animal power both have heavy start-up and shut-down costs.) The new methods usually reduced unit costs substantially, and this has attracted the attention of historians. But, at least equally important for the argument here, and more important in the long run for the nature of the market, was the change in the degree of adaptability of costs.

These pressures also tend to change the nature of competition; previously it centred on prices in a comparatively simple way. But now the chief focus will become technological development, especially in relation to market share, since an increase in share can permit a larger size, which in turn will make it possible to extend the division of labour, reaping economies of scale that in turn will permit the consolidation of a lower price, and so on. Once an advantage is achieved, a 'virtuous' cycle develops, enabling the successful firm to establish a leading or even dominant market position.

A 'virtuous' cycle also develops for the system as a whole. In the craft economy new technical developments are seldom written down and remain tied to particular workers and foremen. These may train others and the techniques will spread by word of mouth, and with the movement of labour. But the diffusion process is subjective, slow and uncertain. By contrast, once mechanization has given birth to a competitive machine-building industry, technical change is on its way to becoming endogenous, for competition in machinery and machine tools will take place chiefly in attempting to make the best improvements; the spread of these through learning and the movement of engineers between firms will begin to make technical change endogenous. As engineering becomes an established profession, it will find a place in universities, together with support from government, while firms will begin to create R&D centres, further reinforcing the position of technical change as a normal aspect of industrial competition.

Yet a major hitch can arise in this: a Catch-22, in fact. It is true that mass production methods provide great advantages, but they also have costs, in particular, large capital costs which cannot be recouped unless there is, in fact, a mass market. Yet a mass market may not develop unless and until the product is widely available at a suitably low price, and without a mass market the capital costs will not be spread over enough units to lower the price sufficiently to make the new methods competitive. This creates the possibility of a bizarre dilemma: it may not be worth investing in further expansion of craft production because mass production methods are being developed, and are superior at high levels of demand. Craft investments will be wiped out as soon as mass production begins. But a mass market does not yet exist, and if any firm starts mass production at least several others will follow; and there will certainly not be room for all until the mass market actually develops, and perhaps not even then. Hence it may be too dangerous to make the large investment. Consequently, nothing will be done and the economy could stagnate for lack of investment.

A mass market is essential, for the huge investment can only be recouped by long production runs. Here a further factor may intervene: as the new mass product enters into wider usage, 'learning by using' may occur. Aeroplanes will fly under new weather conditions, land and take off in different circumstances, and a new knowledge of the strengths and weaknesses of the technical design will become manifest. Customers will demand modifications, and the long production run will be lost (Argyrous, 1991). Frequent design changes and significant learning-by-using are incompatible with a move to mass production.

A further problem arises in that craft work teams are not likely to want to implement a move to mass production. Even when it is clear that moving to such methods would be warranted, craft workers and foremen are likely to resist innovations that will render many of their skills obsolete. Neither will they will find it easy and natural to adapt. (Hounshell, 1984).

There are several ways out of these problems: cartels may be formed and the market parcelled out, or the state may intervene, as it did in the Second World War, to plan and finance the transition. This may be a more or less straightforward matter of providing finance and assuring the market, as in shipbuilding, or it may involve far more detailed and difficult planning, extending to product design and specifications, as in aircraft. So the problem is serious; the very effectiveness of the new methods may act as a barrier to their adoption.

Nevertheless, if and when investment does begin, with competition centring on a race for improvements in technology, a shift to mass production methods will take place (Thomson, 1989). With a little coaxing, perhaps the preceding could be blended into a familiar stylized picture: the famous sigmoid diffusion curve. Plot an index of diffusion say, the percentage of firms using, or of output produced by the new method) on the vertical axis, with time on the horizontal. Then, at first, the Catch-22 will hinder adoption and the curve will be shallow; but once a few firms have made the switch and it is clear that a market is there, competitive pressures will compel the mass of firms to rush to make the change, and the curve will rise steeply. Finally, it will flatten out, because the remaining craft producers will be those holding protected niches in the market, insulating them from competitive pressures.

One result will be that the strategic situation of firms will be fundamentally changed. Firms will no longer seek to establish their optimum size and remain there, and it will no longer be possible to permit new firms to supply the growth of the market. New firms would be able to build plants and buy equipment embodying the latest technology; they would therefore be able to establish a cost advantage and invade the markets of established firms. Hence existing firms must invest regularly, incorporating new technology into their plants, and growing enough so that, taken together, they will supply the expansion of the market. Prices will therefore have to be set with an eye to providing the profits that will finance growth. To facilitate such growth, firms will withhold profits, investing them directly, rather than distributing them to shareholders. Both the pattern of competition and the working of financial markets is altered by the move to mass production technology and the shift from extensive to intensive growth.

Consequently, when the technology of mass production becomes widespread, the system as a whole will begin to work differently. First, when demand for a particular good falls – say, due to a general decline in investment spending – workers will be laid off, and prices will tend to fall comparatively little. Thus instead of consumption varying inversely with investment, as it did when a decline in demand led to a greater proportional fall in prices than in money wages, consumption will also decline, since the laid-off workers and their families will now have to curtail their consumption spending. The elasticity of consumption with respect to investment will now be positive. The dog begins to bark.

Notes

1. Changes of this kind are continuing and the spread of mass production technology in developing countries is evident, but in the advanced nations we can perhaps see the beginnings of a new kind of economy, based on information systems. Real-world economies, of course, always contain mixtures of (imperfect versions of) the idealized systems portrayed in theory.
2. Sraffa contended, first, that moving along a partial equilibrium supply curve would cause the demand curve to shift, unless all or most of the economy's fixed factor is used by (only by) the industry in question; and, second, increasing returns must be external to the firm but internal to the industry, otherwise the size of the firm will be indeterminate. Clearly these are very special cases, and quite implausible for modern industry, but they fit the traditional farming and craft economy quite well. Land is the chief fixed factor in agriculture, and traditional skill, passed on from father to son, is the fixed factor in most crafts. As for increasing returns external to the firm, when agricultural output rises sufficiently in an area, new forms of storage and transport will be introduced (the railways, grain elevators, new roads, etc.). Similarly, when crafts develop to a certain size, bulk buying, bulk storage and improved, marketing can be introduced.
3. The domestic system provides for flexibility in wage incomes, leading to variation in worker consumption; fixed interest income provides stability to capitalist households, enabling them to maintain consumption. If these two approximately offset each other, fluctuations in investment will still be offset by opposite movements in consumption, and the analysis of Nell, 1991, Ch. 16 will continue to hold (see below). These two systems correspond in many respects to Marx's distinction between two paths to the formation of industrial capital. The route of merchants entering production and employing wage-labourers, usually through 'putting out', is termed the more conservative path since it tends to preserve most of the older methods of production. By contrast, the 'really revolutionary way' is for producers (craftsmen) to develop into merchants and capitalists by

expanding their establishments, hiring additional workers and learning to handle their own marketing. This path will lead to innovations and new products, but it also carries greater risks.

4. In practice, of course, there will be dispensable labourers, and many opportunities for partial shut-downs; moreover, the practice of 'putting-out' continued in many industries long after the introduction of the factory system, perhaps precisely because its flexibility in response to demand changes tended to offset its higher unit costs of production. These important historical points will be set to one side, however, because the aim here is to draw the contrasts sharply, basing them on distinctive features of the respective technologies.
5. In the pin factory each man specializes in a particular operation, drawing the wire, straightening it, cutting, pointing, grinding, putting the head on, whitening or polishing, etc., amounting to about 18 distinct operations which can be divided up in several different ways. Clearly they have to be coordinated, and each worker has to complete his operation in a manner that eases the way for the next. The description makes it clear that the division of labour itself requires practice; the advantages cannot be reaped without effort and skill. Moreover, much of the machinery and improvements 'were originally the inventions of common workmen, who, . . . naturally turned their thoughts towards finding out easier and readier methods' (Smith, 1904, p. 11) Marshall argues that while machinery displaces purely manual skill, it increases the demand for judgement and general intelligence. Again his examples illustrate the need for coordination of the workforce and for high morale (cf. pp 255–64).
6. With characteristic acumen Marx observed, 'On a closer examination of the working machine proper, we find in it, as a general rule, though often, no doubt, under very altered forms, the apparatus and tools used by the handicraftsman or manufacturing workman; with this difference, that instead of being human implements, they are the implements of a mechanism, or mechanical implements. Either the entire machine is only a more or less altered mechanical edition of the old handicraft tool, as, for instance, the powerloom, or the working parts fitted in the frame of the machine are old acquaintances, as spindles are in a mule, needles in a stocking loom' (Marx, Vol. I, p. 373).
7. Adam is not the only smith we can cite:

> Under a spreading chestnut tree
> The village smithy stands
> The smith, a mighty man is he
> . . .
> His brow is wet with honest sweat
> He earns what e'er he can
> . . .
> Week in, week out, from morn till night
> You can hear his bellows blow
> You can hear him swing his heavy sledge
> With measured beat and slow.
> (Henry Wadsworth Longfellow)

'He earns whate'er he can': that is, he works steadily, 'week in, week out', although his earnings may vary with demand. That a craftsman works steadily but receives earnings which fluctuate with good times and bad is an idea so natural that it appears almost unconsciously in poetry and novels.

8. The simplest matrix multiplier follows from the linear production equation: $x = Ax + y$, which implies

$$x = [I - A]^{-1}.$$

Here x will be the column vector of gross output, y of net, and A the square, non-negative, irreducible input coefficient matrix. Several variations are possible on this theme, but all canter on the idea that, behind any net output, there lies a 'vertically integrated' sector of inputs: the direct plus indirect labour required, for example, or, calculated in the same way, the direct plus indirect quantities of each of the various means of production. Then, on the assumption of constant returns, an increase or decrease in an output will imply a corresponding increase or decrease in the vertically integrated inputs, and the ratio between these will be the multiplier.

But this is not the multiplier of Keynesian economics, which gives the magnitude of the secondary changes brought about by the short-run adjustments to a given initial change. Vertically integrated inputs can be calculated for *any* interdependent industrial system exhibiting fixed coefficients or constant returns in some sense. For example, in a fixed employment system, the labour and inputs required for a unit of the consumer good can be calculated, and then the labour and inputs required for each of those inputs, and so on. This will sum to a definite magnitude, provided the A matrix is productive, etc. but no short-run variation is possible, as we have seen. In the long run, however, an expansion of production of the consumer good will require the calculated secondary expansion of labour and inputs, and labour and inputs for labour and inputs, etc. This simply reflects the logic of interdependent production, and is important in that context; but it tells us nothing about the pattern of short-term adjustment. This depends on the passing along of expenditures, therefore upon variable costs (and household withdrawals, if any).

As a first approximation, assume that all labour costs are variable; assume also that the means of production of consumer goods must vary with consumer goods output and, when they do, their labour inputs vary. Wages will be paid in units of command over the composite consumer good; all wages will be spent on consumption. Then we have,

$$y = c + i,$$

where c and i are the vectors of consumption and investment. Then $c = wNx$, where w is the real wage, N is the vector of direct labour inputs, and x the vector of gross output. Hence,

$$c = wN\,[I - A]^{-1}y = wL^{*}y,$$

where $L^{*} = N[I - A]^{-1}$: that is, L^{*} is the total direct and indirect labour employed per unit output.
So we have,

$$y = wL^{*}y + i,$$

and, rearranging,

$$y = [I - wL^{*}]^{-1}i,$$

a matrix reformulation of the multiplier described in the text. It depends on the real wage and on productivity; prices do not enter explicitly (as they would have to if, for example, there were saving out of income, or if we had to aggregate different patterns of consumption for different classes). It also depends on the assumptions about variability; if only part of the labour costs are variable, the vector N will have to be adjusted accordingly for the analysis of changes in y corresponding to changes in investment.

9. The failure to appreciate this may be due to a common confusion of two distinctions, between fixed and variable costs, on the one hand, and capital and current costs, on the other. Fixed costs are incurred at a set rate so long as the firm is in business; variable costs are incurred in proportion to the level of output. Current costs are those incurred in the current accounting period; capital costs are spread over many accounting periods. Variable costs are always current, but not all current costs are variable. Labour costs are current; but in the craft economy (and in part, in mass production, too) labour may be fixed. Capital charges are not necessarily fixed. Depreciation is a capital charge, but it may be varied with output; interest and dividend payments may be varied with earnings.
10. The modern textbook short-run production function is a comparative-static construction which shows the different outputs resulting from various amounts of labour combined with a fixed amount of *capital*, not a fixed set of capital goods. In the modern version the given amount of capital, or funds, will be embodied in the most appropriate form at each point, and will always be fully utilized. But, speaking of what he called the 'fixed stock of appliances for production', 'when the supply is excessive (in relation to current demand),' Marshall remarks 'some of them must remain imperfectly employed', and when the 'supply is deficient, . . . producers have to adjust . . . as best they can with the appliances . . . at their disposal'. Clearly the idea is that the labour force varies the intensity with which it operates the fixed set of appliances and, as the intensity of utilization varies, so will productivity. By contrast, in an industrial economy, when utilization varies, *employment* changes.
11. If the multiplier is m, then $C = mI$, and $dC = mdI$; hence, rearranging, $IdC/CdI = +1$, the elasticity of consumption spending with respect to investment is unity. The proportional changes in the two components of demand are equal (Nell, 1991, Ch. 16).
12. Marshall clearly understood that prices and money-wages had different degrees of flexibility, though he largely failed to draw out the consequences:

> when prices are rising, the rise in the price of the finished commodity is generally more rapid than that in the price of the raw material, always more rapid than that in the price of labour; and when prices are falling, the fall in the price of the finished commodity is generally more rapid than that in the price of the raw material, always more rapid than that in the price of labour.

Even though he seems to have reversed the actual relationship between the variability of the prices of finished goods and raw materials (Pedersen and Petersen, 1938; Ashton, 1948), he grasped the main point: 'statistics prove that the real income of the country is not very much less in the present time of low prices, than it was in the period of high prices that went before it. The total amount of necessaries, comforts and luxuries which are enjoyed by Englishmen is but little less in 1879 than it was in 1872' (Marshall, 1961, Vol. 2, pp. 714–16).

Kaldor, in 1938, clearly understood that if prices were flexible relative to money wages, the resulting distribution effects would bring a Wicksellian 'cumulative process' to an end (Kaldor, 1960, pp. 110–11).

13. It can be shown that, in Artisan economies, given somewhat special conditions, the elasticity of consumption with respect to investment will be -1. More generally, under plausible assumptions, it will be negative, between 0 and -1 (Nell, 1991, Ch. 16).
14. The classical price equation, for single-product industries and circulating capital – the simplest case – is:

$$p = (1 + r)Ap + wL,$$

where p is the relative price vector, r the rate of profit, w the real wage, A the input output matrix and L the vector of labour inputs. These are long-run prices – Marshall's case of the horizontal supply curve – raising the question of what happens in the short or medium term. An important concern of contemporary economics is the extent to which, under various circumstances, prices are flexible in the face of variations in demand, a short or medium term question. Hicks (1939, 1965, 1977) distinguished 'fixprice' and 'flexprice' economies, and corresponding to the different systems, fixprice and flexprice methods of analysis, Value and Capital being an example of the latter, Capital and Growth of the former. But he provided no systematic explanation of why prices should be flexible in some circumstances and rigid in others. Sylos-Labini argued that the explanation lay in the growth of market power and methods of organizing and controlling markets (1984). Yet 'competition among the few' is surely intense, and the question is why does it not affect prices more? More recently, Okun distinguished 'customer' and 'auction' markets, the former exhibiting fixed or managed prices, while the latter's prices are flexible and market-driven. But why the growth of 'customer markets', and why do specific markets take one form or the other? Since Okun a large number of 'new-Keynesian' writers have elaborated the distinction and proposed further reasons to explain the inflexibility of prices in modern economies in the face of fluctuations in demand (Gordon, 1990). The fact of comparative inflexibility is unquestionable, but none of these explanations to date has commanded general assent.
15. The blocking of entry here is the result of an investment strategy, not of a pricing strategy. A considerable literature has developed around the idea that oligopolists would fix on a price lying above average costs, but still below the level that would induce (or permit) entry, or would threaten to cut to such a price on perceiving preparations to enter (Sylos-Labini, 1969; Modigliani, 1958; Bhagwati, 1970; Wenders, 1967). Much of this literature is still wedded to the idea that prices are determined in conjunction with output, thereby missing the connection with investment. In any case, the idea depends crucially on the existence of barriers that permit a gap between average costs and the entry-inducing price. In the present analysis no such barriers are assumed; markets are competitive and entry is always possible, but those already in the market do have the advantages of experience and established connections. Entry is blocked because the existing firms have the capacity and are already in position to supply the entire potential market.
16. If equilibrium is conceived as a *position*, it will be undermined by technical change. But if it is thought of as a process or a path, it may well be determinate (Goodwin, in Halevi, Laibman and Nell, 1991; Nell and Semmler, 'Introduction', in Nell and Semmler, 1991).

17. Elsewhere I have argued that a corporate industrial economy is normally constrained by its level of demand, rather than by resources, capacity or labour. Further, since demand is scarce in every field, competition sets up pressures for cutting costs and raising productivity, with the result that capacity output is increased and the multiplier reduced, so that the gap between aggregate demand and potential supply is always tending to widen. Cf. Nell and Semmler, 1991, Ch. 31, 'Capitalism, Socialism and Effective Demand'.
18. In a perfect capital market, any borrower should be able to raise any amount at the prevailing rate of interest. A corollary is that the roles of worker and employer are adopted by choice; labour can hire capital just as easily as capital can hire labour. Here as elsewhere the ideas sketched by Marshall in the context of an artisan economy come to grief when applied to industrial conditions. A 'perfect capital market' cannot account for the financial structure of a modern corporation. Consider a bank or a group of investors being offered a bond by an entrepreneur. The security will be the project itself, and the return the prevailing rate of interest. Both sides must consider the advantages and disadvantages of the actual project. For the entrepreneur the choice is between borrowing capital and owning the project and so working for himself, or working as a manager for another owner. So the expected profit from the project must exceed the required interest plus the competitive salary for managers. The bank/investor group faces the analogous choice. They can lend to the entrepreneur or hire a manager and own the project directly. If the expected profit from the project exceeds the going interest plus the competitive manager's salary, they should set up the project themselves and hire a manager. If the expected profit falls short of interest plus salary, they would not want to own it but, by the same token, they should not accept the assets of the project as security either, since the income it generates is not sufficient to meet its costs. So when a project is potentially profitable both investors and entrepreneurs will wish to hold the equity, and when it is not, investors should not consider it adequate security. The assumptions have precluded any explanation of financial structure, and this is what lies behind the Modigliani–Miller theorem. (It could be argued that entrepreneurs and investors have different knowledge and skills, and therefore different expectations of profitability. This is sensible enough, but in a 'perfect' market there has to be perfect information. Moreover, if there is imperfect knowledge, there is a good motive for requiring independent security for loans, in which case borrowing power will be limited by net worth.)
19. These improvements may change the nature of the goods or the techniques. It is therefore not possible, in the case of an industrial economy, to start with the assumption that the list of products and the set of techniques is given at the outset. On the contrary, firms will compete to produce new and improved products and cheaper techniques. To assume a fixed list, however, large, implies ruling out the form of industrial competition that determines the nature of both labour and the investment process. Or, to put it another way, the assumption of a fixed list is an essential underpinning of the idea that prices and quantities are determined simultaneously by market-clearing.
20. In early or fixed employment capitalism an appropriate simplification would be to consider the two classes of capitalist families, the owners of firms and worker families. But in advanced or flexible employment capitalism, capital is no longer lodged with families; it is held in professionally managed portfolios and/or by financial institutions. The former owners are now beneficiaries or largely passive recipients of interest and dividends; their interests in life centre around their careers. (Cf. Nell, 1990)

21. A fuller and more sophisticated treatment would consider the possibilities of trade-offs between the characteristics composing the lifestyles. This must be done carefully: nutrition and taste may trade off in diet, and elegance and comfort may trade off in clothing or shelter. The possibilities of substitution are far fewer, however, when it comes to the broader categories of diet, clothing and shelter: a smaller house for better clothing? Eat cheaply to dress smartly? On occasion, no doubt; but as a general rule a certain standard will have to be achieved in each category, but the significant trade-offs will be in the way it can be reached. For our purposes, however, these questions can be set aside. (For further discussion, see Nell, 1991, Ch. 17).

References

Argyrous, George (1991) 'Investment, Demand and Technological Change: Transformational Growth and the State in America during World War II', PhD Dissertation, New School For Social Research, New York.

Broehl, Wayne, (1959) *Precision Valley: The Machine Tool Companies of Springfield, Vermont* (Englewood Cliffs, NJ: Prentice-Hall).

Ashton, T.S. (1948) *The Industrial Revolution, 1760–1830* (New York: Oxford University Press).

Barnes, R.M. (1956) *Motion and Time Study* (New York: John Wiley).

Bhagwati, Jagdish (1970) 'Oligopoly Theory, Entry Prevention, and Growth', *Oxford Economic Papers*, 22, pp. 297–310.

Chandler, Alfred (1977) *The Visible Hand* (Cambridge, Mass.: Belknap Press).

—— (1990) *Scale and Scope* (Cambridge, Mass.: Belknap Press).

Eatwell, J. and Milgate, M. (1983) *Keynes' Economics and the Theory of Value and Distribution* (New York: Oxford University Press).

Eichner, Alfred (1976) *The Mega-Corp and Oligopoly* (New York: Cambridge University Press).

Garegnani, P. (1983) 'On a Change in the Notion of Equilibrium in Recent Work on Value and Distribution' in Eatwell and Milgate (1983).

Gordon, Robert (1990) 'What is New-Keynesian Economics?', *Journal of Economic Literature*, 28, 3 (Sept.).

Halevi, J., Laibman, D. and Nell, E. (eds) (1991) *Beyond the Steady State* (London: Macmillan).

Hicks, Sir John (1939) *Value and Capital* (New York: Oxford University Press).

—— (1965) *Capital and Growth* (New York: Oxford University Press).

—— (1977) *Economic Perspectives* (New York: Oxford University Press).

Hounshell, David (1984) *From the American System to Mass Production, 1800–1932* (Baltimore, Md.: Johns Hopkins University Press).

Howell, David (1991) 'Stages of Technical Advance, Industrial Segmentation and Employment' in Nell (ed.), *Economics in Historical Perspective: Essays in Worldly Philosophy* (London: Macmillan).

Kaldor, Nicholas (1960) *Essays on Value and Distribution* (London: Duckworth).

Landes, David (1972) *The Unbound Prometheus* (Cambridge University Press).

Leijonhufvud, Axel (1986) 'Capitalism and the Factory System', in Richard Langlois (ed.), *Economics as a Process* (Cambridge University Press).

Maddison, Angus (1982) *Phases of Capitalist Development* (New York: Oxford University Press).

Mantoux, Paul (1928) *The Industrial Revolution in the Eighteenth Century* (London: Methuen).
Marshall, Alfred (1961) *Principles of Economics*, 9th (Variorum) edn (London: Macmillan).
Marx, Karl (n.d.) *Capital* (New York: International).
Modigliani, Franco (1958) 'New Developments on the Oligopoly Front', *Journal of Political Economy*, 64, pp. 215–33.
Nell, Edward (1990) 'The Institutionalization of Capital' (New York: New School, unpublished manuscript).
—— (1991) *Transformational Growth and Effective Demand* (London: Macmillan).
—— and Semmler, W. (1991) *Nicholas Kaldor and Mainstream Economics* (New York: St Martin's).
Ochoa, E. (1984) 'Labor Values and Prices of Production: An Interindustry Study of the U.S. Economy, 1947–72', PhD Dissertation, New School for Social Research, New York.
Pederson, J. and Peterson, O. (1938) *An Analysis of Price Behaviour* (London: Humphrey Milford and Oxford University Press).
Robinson, Joan (1980) *Further Contributions to Modern Economics* (Oxford: Basil Blackwell).
Sabel, C. and Zeitlin, J. (1985) 'Historical Alternatives to Mass Production: Politics, Markets and Technology in 19th C Industrialization', *Past and Present*, 108.
Smith, Adam (1904) *An Inquiry into the Nature and Causes of the Wealth of Nations* (London: Methuen).
Sraffa, Piero (1926) 'The Laws of Returns Under Competitive Conditions', *Economic Journal*, 36.
Sylos-Labini, P. (1969) *Oligopoly and Technical Progress* (Cambridge, Mass.: Harvard University Press).
Sylos-Labini, P. (1984) *The Forces of Economic Growth and Decline* (Cambridge, Mass.: MIT Press).
Thomson, Ross (1989) *The Path to Mechanized Shoe Production in the United States* (Chapel Hill, NC: University of North Carolina Press).
Wenders, J. (1967) 'Entry and Monopolistic Pricing', *Journal of Political Economy* 75, pp. 755–62.
Wood, Adrian (1976) *A Theory of Profit* (Cambridge University Press).

13 Borrowing Technology or Innovating: An Exploration of the Two Paths to Industrial Development

Alice H. Amsden and Takashi Hikino

Great Britain, and then the USA and Germany, became world industrial leaders by generating pioneering technology. They either invented new products and processes or were the first to commercialize them on a large scale. By contrast, late-industrializing countries in the twentieth century have evolved as 'learners', by borrowing and improving technology that had already been commercialized by experienced firms from more advanced economies.

While innovators in the First and Second Industrial Revolutions certainly borrowed and learned from each other, even the most prominent enterprises in late-industrializing countries have had to grow exclusively on the basis of learning; they initially had no competitive assets in a new product or process, which is the meaning we attribute to 'lateness'. The imperative to industrialize exclusively on the basis of learning is responsible for many general properties in common in a subset of developing countries that are diverse in resource endowment and culture: Japan (although as a pioneering learner, unique in many respects), South Korea, Taiwan, India, Brazil, Mexico and others (Amsden, 1989). It is these countries that generally have increased dramatically their share of world production in the twentieth century (see Table 13.1).

On the one hand, standard price theory predicts that latecomers develop by 'getting the prices right' and typically using low wages to gain a comparative advantage in labour-intensive industries. But, as discussed shortly, no major late industrializing country has successfully followed this route in the twentieth century. On the other hand, the doyen of institutional theories of late industrialization, Alexander Gerschenkron, conceives catching up as a process of 'revolutionary', 'eruptive' spurts, with the most backward countries promoting 'those branches of industrial activities in which recent technological progress has been particularly rapid' (1962, pp. 9–10). Not, however, since the now discredited case of Russia, whose experience principally Gerschenkron sought to explain, have latecomers followed this route.[1]

Table 13.1 Distribution of world GDP, 1900–87*

Country	*Year (% of 32-country total)*				
	1900	*1913*	*1950*	*1973*	*1987*
North America and Western Europe†	54.0	57.5	61.7	53.5	47.0
UK	10.4	9.1	7.1	4.8	3.8
USA	21.5	25.7	34.6	26.9	24.3
Germany	5.1	5.4	4.2	5.4	4.4
Japan	2.9	2.9	3.2	8.3	8.8
South Korea‡ and Taiwan	0.6	0.5	0.5	1.1	2.0
Brazil and Mexico	1.6	1.7	3.0	4.4	5.1
India	8.6	7.0	4.4	3.4	3.8

* 1980 prices, adjusted to exclude impact of boundary changes.
† Austria, Belgium, Canada, Denmark, Finland, France, Germany, Italy, Netherlands, Norway, Sweden, Switzerland, UK and USA.
‡ Rough estimate for 1900.

Source: Computed from Maddison (1989), p. 113.

Instead, the leading enterprises of the late industrialization spearheaded by Japan have pursued a strategy of competing primarily by means of incremental rather than radical changes in products and production techniques. They have entered those branches of industry, whether capital- or labour-intensive, that are not new but rather 'post-adolescent' or 'mid-tech'. A 'post-adolescent' technology is one that can be purchased at competitively-determined prices in the international market, much like many other products.[2]

It is our intention in this chapter to present some of the common properties of a historical paradigm of industrialization through learning. We present a theoretical description of industrializing through borrowing technology that is set in a specific historical context, the twentieth century. We use the process of technology acquisition – structured in terms of either generating new, or borrowing post-adolescent technology – as a cutting knife to distinguish the economic behaviour of countries in the course of industrializing. By comparing the historical industrialization process of borrowers and innovators, we argue that technology acquisition differences strongly shape: (a) the *role of government* in promoting industrial development; (b) the *competitive focus* of successful leading enterprises; and (c) *firm strategy and structure*.

A HISTORICAL PARADIGM OF LEARNING

The long-term success of industrialization depends on achieving international manufacturing efficiency, either by competing against imports or by capturing export markets. We will stress the latter given the relatively small domestic market from which most twentieth-century countries began to industrialize, and given the scale rigidities of certain imported technologies with respect to down-sizing. In other words, in the case of post-adolescent technologies where an internationally competitive scale of operation cannot be altered and exceeds domestic absorptive capacity, exports become the key to the productive efficiency of learners.

It is easier for a country to industrialize the larger its bundle of competitive assets, where assets are defined to consist of anything that adds to the international competitiveness of raw labour power (say, natural resources, labour skills, managerial expertise, physical capital and, of course, proprietary technology in the form of new products and processes). The greater a country's assets, the higher its value added per worker (Lary, 1968). Raw materials and artisan skills have assisted capital accumulation in many countries, but virtually no late industrializer has prospered on the basis of raw materials or artisan skills alone.[3] Therefore we do not discuss such endowments, and assume that the major asset which late-industrializing twentieth-century economies bring to bear in world markets is their cheap labour.

At best, however, low wages can be used as a 'cash cow' but cannot serve any better than raw materials or artisan skills as the dynamic engine of modern economic growth in the long-term; as wages tend to rise in the course of industrialization, labour-intensive industries promptly lose competitiveness to still lower-wage countries. Furthermore, even labour-intensive industries may not serve as cash cows because low wages alone may be an inadequate competitive weapon against the higher productivity levels of more advanced countries. Standard price theory assumes identical production functions in the same industry in all countries, but in reality more advanced economies may be more cost-effective even in industries with high labour content due to superior infrastructure in their operating setting, their better managements and workforce skills, as well as their cache of tacit, non-transferable productivity and quality improvements. The standard form of technology transfer – designs, blueprints and production equipment – or even a turnkey transfer, is usually insufficient to overcome the productivity gap.

Under such circumstances, exchange-rate devaluations may lower real wage costs in international markets, but currency depreciations can only occur within limits set by politics, workers' physiological intake requirements and the need to import production inputs. Even after repeated real currency devaluations in twentieth-century industrialization, low wages have been found to be an insufficient competitive advantage in world markets, notwithstanding the labour intensity of a leading sector like cotton textiles (see, for example, G. Clark, 1987).

As discussed in detail later, therefore, a *necessary condition to industrialize in the twentieth century is systematic government intervention*. Given the inadequacy of low wages and the absence of pioneering products and processes, governments must intervene and distort prices to get them 'wrong' rather than right, make subsidies contingent on the achievement of performance standards, and monitor subsidy recipients to ensure that productivity, and not just profitability, rises (Amsden, 1989; 1991a; 1992).

Needless to say, the modern theory of government intervention in late industrialization originated in Gerschenkron's ideas on backwardness in European history (see Fishlow, 1987, for a concise summary). Gerschenkron viewed the increase in government intervention in sequentially later industrializations as a consequence of the shortcomings of institutional actors in the marketplace, particularly the capital market, rather than as a consequence of the market mechanism itself failing to generate industrial growth. Yet government intervention cannot simply take the passive form of providing remedies for institutional shortcomings or even market failures; it must proactively intervene in the determination of firm's production costs. The need for such intervention is greater in industries where the competitive asset bundle of advanced economies is greater.[4]

If late industrializers cannot prosper by means of low wages alone, neither can they compete by developing high-tech products. In the dynamic industries that are close to the world technological frontier in which the advanced countries operate, the basic rules of international competition changed in the twentieth century as global enterprises with 'organizational capabilities' based on a core technology arose (Chandler, 1990). The institutionalization of R&D in such enterprises allowed them to erect entry barriers around their proprietary technology family. Because of this historical condition, Gerschenkron's idea of leaping to the world technological frontier could no longer work. Earlier, in the nineteenth century, when British firms could not establish equally impenetrable international entry barriers, leading American and German enterprises could and did follow a Gerschenkron-like competitive strategy: they leap-frogged ahead of England in the most dynamic sectors. In the twentieth century, the only major country to attempt this strategy was Russia and, despite partial and purely technical success, it has failed to become a stable, industrialized economy.

With the non-dynamic bottom end of the market vulnerable to lower-wage competitors, and the top end impenetrable due to technology entry barriers, successful late-industrializing countries in the twentieth century have targeted as their dynamic industrializing core post-adolescent or mid-tech industries, where technology, although not free, is available from international suppliers and global demand is growing (in some cases, such as steel, simply because world population is growing). Other examples, in historical sequence, are electrical machinery, basic chemicals, motor vehicles, consumer electronics and commodity semiconductors.

Nevertheless, even if investments in mid-tech are the correct allocative choice necessary to industrialize, they are no guarantee of international competitiveness. In order to compete against the post-adolescent products from more advanced economies which enjoy higher productivity and lower cost, late industrializers in the twentieth century must sharpen their own managerial and organizational skills, shorten their learning period and, above all, make incremental improvements in the quality and performance of their process and product. *The shop floor becomes their strategic battleground*. It is initially here where they must make borrowed technology work, even if it comes in the form of a 'turn-key' transfer, because no technology, however mature, is fully documented, completely understood, and therefore perfectly transferable (Nelson, 1987; Landes, 1992).[5] It is on the shop floor where they must adapt borrowed technology to suit their targeted market size and other idiosyncratic conditions. It is also here where they subject borrowed technology to continuous incremental upgrading.[6]

Going one step further, even if a firm successfully identifies a mid-tech industry as the proper one to enter, and even if it invests successfully in incremental product and process improvements, it faces grave uncertainties about the success of its product in international markets. These uncertainties arise because global oligopolies may come up with a revolutionary product that may render a mid-tech product, however seasoned on the shop floor, obsolete overnight. Insofar as the late-industrializing firm is unable to protect itself by innovating further around a core technology family, it protects itself by routinizing a strategy of wide diversification into many technologically-unrelated mature product markets, as in the case of the Japanese *zaibatsu*, Korean *chaebol* and Latin American groups. *The firm structure that results from this strategy is thus a network or collection of technologically isolated firms*. As detailed and systematic knowledge of products and processes is not embodied in the top of this type of group firm, top management tends to confine itself to the functions of resource allocation and monitoring.

We now examine in more detail how the imperative to industrialize by borrowing technology generates each of the three characteristics noted above: government intervention, shop floor focus and group firm structure.

GOVERNMENT INTERVENTION

Even in the classic case of the First Industrial Revolution in England, the government's role never took the pure form of *laissez-faire* (Coats, 1971; Taylor, 1972). Rather, from the late eighteenth to the late nineteenth century the pattern, if not the intensity, was similar in most industrializing countries, mainly taking the form of indirect support for economic development. Britain, the USA and much of continental Europe witnessed the rise of developmental states that tried to create the institutional and physical infrastructure for smoother and faster industrialization. Governments were actively involved in developing transportation, communications and all sorts of education, as well as stable banking systems and legal and administrative frameworks generally. More direct intervention, affecting price competition, increased over time to the extent that tariff protection to infant industries became widespread, for a host of reasons related to revenue and politics, as well as economic development (Goodrich, 1967; Landes, 1969; Nye, 1991).

Developmental states in the twentieth century did all this but much more, examples being Brazil, Turkey, India, South Korea, Taiwan and even Japan; but, because Japan was relatively less underdeveloped, its state could do less than in these other countries. Governments intervened more institutionally because the roadmap of earlier industrializations already existed.[7] The promising industries to promote could be projected in five-year indicative plans. Special ministries responsible for economic development were established. Development banks were formed. The macroeconomic tools of the Keynesian revolution also involved governments in short- and medium-term stabilization exercises. In addition, governments had to subsidize business more because, without pioneering technology, late industrializers had fewer assets with which to compete in international markets.

The rising institutional role of government in late industrialization is evident at the turn of the twentieth century in the function of export promotion, which simultaneously became an issue for various reasons in the USA, Germany and Japan. In the case of the USA, Becker (1982) points out that the commencement of 'foreign sales was achieved for the most part without assistance from the U.S. government' (p. 50), possibly because by that time industry was relatively well developed. But, even in the case of smaller firms (which were less able to market overseas and which sought government assistance), 'those who turned to consuls found much to criticize in the consular service's performance' (p. 92). Germany also relied on the private sector to boost its exports, although a larger role was played by business associations. In the case of Japan, which was just starting to industrialize in the 1890s, the government pushed businesses to organize themselves into associations. Moreover, 'the government eagerly developed policies for the promotion of foreign trade. Above all, the activity of Japanese consulates abroad was remarkable' (Miyamoto, 1988, p. 14).

Well after the Second World War, when Japanese industry may have reached a stage of development comparable to that which American industry reached in the 1890s, the Japanese government continued to exhibit the character of a developmental state (Johnson, 1982, 1988; McCraw, 1986). As even their names suggest, Japan's Ministry of International Trade and Industry, or MITI (renamed in 1949, and originally created in 1881 as the Ministry of Agriculture and Commerce), played a larger role in promoting exports than the role ever played by the US Department of Commerce (founded in 1903 as the Department of Commerce and Labor). The role of the government in export promotion increased still further with later industrializers. The South Korean government set export targets for specific industries and firms, provided exporters with subsidized working capital and made protection of the home market contingent on export success (Rhee, Ross-Larson and Pursell, 1984; Amsden, 1989). By the 1970s and 1980s, the export promotion of the Turkish, Brazilian and Mexican governments had become almost as comprehensive.

The rise in incentives to foster industrial competitiveness during the course of the twentieth century represented more of a departure from past practice than the strengthening of institutional supports. Incentives were possibly least and simplest in Japan, the pioneering learner, and then rose further in still later-industrializing countries which had fewer assets to compete with than Japan, and the added burden of having to compete against Japan itself.

A study in the 1930s to ascertain why the Japanese textile industry was bankrupting Lancashire concluded that Japan's lower wage rates were not responsible (Hubbard, 1938; see also Lazonick, 1990, Ch. 5). Despite Japan's 'beggar thy neighbour' exchange rate devaluations, wages were discounted as the critical variable because it was found that young, female labour in Britain's segmented labour markets earned wages that were not much higher than those being paid in Japan's textile sector. It was, however, unnecessary under the circumstances for the Japanese government to intervene because Japan had competitive assets other than low wages in its arsenal. A superiority over Lancashire derived from Japan's more modern and integrated production facilities (theoretically, capital-rich Britain's competitive advantage), its cartelized bulk-purchase of raw cotton, its better distribution channels and superior management.

By contrast, neither China in the 1930s nor Taiwan and South Korea in the 1960s had anything other than low wages with which to compete against Japan, and low wages were no match for Japan's higher productivity (and its headstart in diversifying into supportive industries, such as synthetic fibres in the case of textiles). An inability to compete on the basis of low wages obtained even after South Korea and Taiwan borrowed foreign technology and devalued their currencies to the point where American aid administrators were satisfied that they had got its price 'right'. To gain competitiveness in cotton textiles, and *a fortiori* in more capital-intensive mature industries, even the governments of these exceptionally low-wage countries (with aid-financed infrastructure) had to intervene to get the prices 'wrong' by providing credit to targeted industries and firms at negative real interest rates, and by distorting exchange rates with tariff protection and incentives to export.

South Korea and Taiwan did not grow faster than other late-industrializing countries by bowing deeper at the altar of free markets, but by allocating subsidies according to different principles. Subsidies were allocated to business according to the principle of *reciprocity*, conditional on business meeting concrete performance standards such as output and export targets, which were then monitored. By contrast, where subsidies were allocated according to the principle of 'giveaway', growth tended to be slower. The state in all late-industrializing countries sternly disciplined labour, but in the faster-growing ones it also disciplined capital (Amsden, 1989, 1991a, 1992). While the rise of government intervention over time has been widely recognized, the reasons for its rise have not always been understood, and the reasons for the variance in its quality across countries have tended to be obscure.

That the discipline of business by government has strengthened over time is shown by the machine tool industry, an interesting case because government intervention in this industry even in the twentieth century has tended to be below the norm. In Japan's machine tool industry, the government's attempts to rationalize production and merge small-scale enterprises in the 1950s were failures (Friedman, 1988). Nevertheless, MITI did succeed in imposing one standard type of numerical control (NC) on all Japanese machine tool builders in the 1970s, which was considered a key factor behind the wide diffusion of NC technology to small users (March, 1989). In the USA, the military wanted the standardization of machine tools to hasten production in case of war but, by contrast, failed to impose its will on the industry (Wagoner, 1968, p. 114). Moreover, indirect subsidies from the military as well as intermittent tariff protection were not tied to performance standards in the USA. On the other hand, tariff protection in the Japanese machine tool industry was differentiated to favour production only of certain machine tool types, and tied to a target of 50 per cent NC output by the end of the 1970s (a goal reached in 1983: Amsden, 1991b). Although government support was negligible in the Taiwan machine tool industry's early growth phase (Amsden, 1977, 1985), it increased later to help the industry ratchet up into a higher quality niche and acquire financially troubled American machine tool companies (OECD, 1990). Preferential treatment became contingent on firms increasing their training and R&D (Dahlman and Sananikone, 1990).

COMPETITIVE FOCUS

If the technologies of the First Industrial Revolution were relatively simple and inexpensive to apply (Musson, 1972), by the late nineteenth century the commercial application of new technologies had become much more complicated and costly. Often, therefore, inventors did not successfully apply or commercialize their own discoveries. In Germany, and particularly in the USA, entrepreneurs began to exploit new technology devised either by themselves or by others (Hughes, 1989). They did so by establishing large organizations that could systematically utilize and further develop the potential of new technologies.

Therefore the strategic focus of the leading enterprises of the Second Industrial Revolution became *top management organization*, with its proprietary asset of technologically experienced managers, controlling large-scale plants and distribution networks. Once founded, these organizations could exploit new technology, increasingly in the context of R&D laboratories.[8] On the basis of these assets or organizational capabilities, first movers established themselves as members of international oligopolies, which made it difficult for newcomers to up-stage them in Gerschenkronian fashion (Chandler, 1990).

In response, late-industrializers in the twentieth century made incremental shop-floor improvements of existing products their principal competitive weapon. Of course, leading enterprises in the First and Second Industrial

Revolutions did not ignore the shop-floor or fail to invest in incremental change; the best of them clearly did (Rosenberg, 1982, ch. 3). But in the First, and especially in the Second, Industrial Revolutions what happened on the shop floor was a *reaction* to the function of new technology generation at the top of the organization. In the nineteenth century the shop floor followed the firm's expansionary strategy, whereas in the twentieth century it led it, as indicated by an example from steelmaking.

Steelmaking

Carnegie Steel's top management, and indeed Andrew Carnegie himself, were fanatical about reducing both direct and indirect production costs, especially in the 1870s and 1880s when the trend in the general price level worldwide was downward. As Carnegie's biographer notes: 'Costs would always be Carnegie's obsession in business, and his constant concern to reduce them in every department was to a large measure the secret of his success' (Wall, 1970, p. 337). What is different in the histories of early and late steel-making countries is that such cost reduction in the former was pursued in conjunction with a more fundamental quest to lower costs by means of major technological breakthroughs, not least of all the costs associated with a high-wage economy. As David Brody writes in his history of American steelworkers: 'The impulse for economy shaped American steel manufacture. It inspired the inventiveness that mechanized the productive operations' (1960, p. 2).

Innovation and investments in incremental improvements operated in conjunction in the sense that Carnegie's search for ways to reduce production costs was a matter of trial and error. He learned from other steelmakers, but he was an innovator to the extent that he had no clear model to follow. For example:

> Carnegie's insistence upon trying out the open hearth process in the mid-1870s . . . proved to be wise. The Siemens furnaces not only enabled the company to produce a higher grade of steel for special orders, but they also proved to be a valuable laboratory for experimentation that would eventually prove the practicality of the open hearth system for the mass production of steel. *Carnegie's introduction of the Siemens process at this early date gives further evidence that he did not follow his own dictum, 'Pioneering don't pay'*. (Wall, 1970, p. 321; italics added)

The absence of a model to follow made Carnegie a pioneer, whereas subsequent steelmakers had Carnegie as a model to follow. In the case of the Brazilian USIMINAS steel mill, for example:

> A market crisis forced USIMINAS to stretch the capacity of its original equipment in order to get a better capital output ratio. Such capacity stretching was possible thanks to the implementation of a standard cost system with an elaborate organizational infrastructure to study its existing

> equipment, *compare it to the best world performance, and then try to reach the same or higher levels*. (Dahlman and Valadares Fonseca, 1987, p. 163; italics added)

As followers, the steel industries of Japan, South Korea, India, Brazil, and other late-industrializing countries had a model to follow as well as technology to buy, but even when technology was acquired through foreign license rather than reverse engineering, it did not get transferred in complete working order. As Richard Nelson (1987) points out, technology is implicit and tacit rather than fully specified and thoroughly understood. It must be adapted and modified in a new environment in order to work at 'rated' capacity, if at all. Moreover, cost constraints forced companies like USIMINAS to employ inexperienced local talent to grope towards rated capacity and beyond, while a first-mover like Carnegie Steel employed the best engineering talent (Alexander L. Holley, who supervised the building of Carnegie's steel works, was considered 'the greatest authority on Bessemer steel mills in America, if not the world': Wall, 1970, p. 312).

South Korea's highly efficient state-owned steel company (POSCO for short), whose survival depended on how well technology from the Nippon Steel Corporation of Japan was absorbed and improved, first assigned its best people to work with foreign plant designers and engineers, and then assigned them to work on 'line' rather than 'staff' functions. This set a precedent of taking production seriously, and it was reinforced by requiring all managers to serve a term on the shop floor as part of their recruitment training. Immediately after commencing operations POSCO faced a situation of excess demand for steel in its fast-growing domestic market. Its objective became increasing its output volume, which it did by using some of the technical skills it had acquired in the technology transfer experience, such as know-how related to downtime minimization, the stabilization of operations, and the optimization of each piece of equipment. POSCO also invested in new capital equipment, but held down engineering costs by using its own staff with experience in the first phase of technology transfer to substitute for foreign technical assistance in some types of capacity additions in the second phase (Amsden, 1989).

A similar regard for nurturing a dedicated corps of production-oriented managers and workers is evident in TISCO, India's integrated iron and steel company, the largest subsidiary of the Tata business group. According to Lall:

> Over the 70-plus years of its existence it has built up a team of singularly dedicated professional managers and technologists as well as a cohesive and skilled labour force which have enabled it to continue with carefully-nurtured but ancient plant. *Much of its survival in adversity can be traced to the technological capability of this large group*. (1987, p. 93; italics added)

Thus, the company's very survival depended on plant- or lower-level management. Lall then goes on to discuss the 'innumerable process improvements' made over time by this group, and by the R&D department, 'which has also worked on process developments' (pp. 96–7).

With respect to R&D of Carnegie Steel's successor, the United States Steel Corporation, it was undertaken very belatedly and then only half-heartedly. US Steel also did not invest aggressively in foreign markets (Chandler, 1990, pp. 138–9). Therefore, when late-industrializing competitors began to emerge on the scene in the 1960s, US Steel could not fight them using either of these competitive weapons, or superior shop-floor capability.

The shop-floor orientation of late-industrializing countries did not preclude their investing in R&D (finance permitting). In absolute value, R&D investments in the steel industry in Japan exceeded those in the US by the early 1970s (Aylen, 1985). In the case of South Korea, POSCO began producing steel in 1973, almost exactly 100 years after Carnegie Steel started production, but whereas over 50 years elapsed from the time of Carnegie's start-up to the time when US Steel opened an R&D laboratory, POSCO made a large investment in R&D only five years after its start-up. Nevertheless, POSCO's R&D, like that of TISCO, was related principally to solving production problems or supporting process engineering. Writing in 1974, Terutomo Ozawa observed that most R&D in a wide range of industries in Japan at that time was also principally production-related. Moreover, even as Japan's competitive focus began to shift by the 1980s from the shop floor to the design office and R&D laboratory, its historical shop-floor orientation coloured the way design and R&D were undertaken. This orientation was evident in a tighter integration of R&D and shop-floor activities than in the USA (R. Hayes and Wheelwright, 1984; R.B. Clark and Fujimoto, 1991; Dertouzos, Lester and Solow, 1989).

GROWTH STRATEGY AND GROUP STRUCTURE

Without a proprietary technology family to protect it and provide it with seeds to grow, the late-industrializing firm is in a precarious position. It is squeezed from below by lower-wage countries and from above by product innovators. It is also threatened by other firms in its domestic market that are following a similar strategy of incrementally improving production techniques. In order to reduce the risks associated with a single product line and overcome the penalty of not having a core technology, firms in late-industrializing countries have diversified widely into unrelated industries, even at an early phase of their corporate development.

Business Groups

Business groups that are widely diversified have acted as agents of industrialization not only in Japan, whose *zaibatsu* became well-known, but also in Brazil, Mexico, Argentina, Peru, Turkey, India, Japan, South Korea, Taiwan, Malaysia, the Philippines, Thailand, South Africa and elsewhere (Okochi and Yasuoka, 1984).

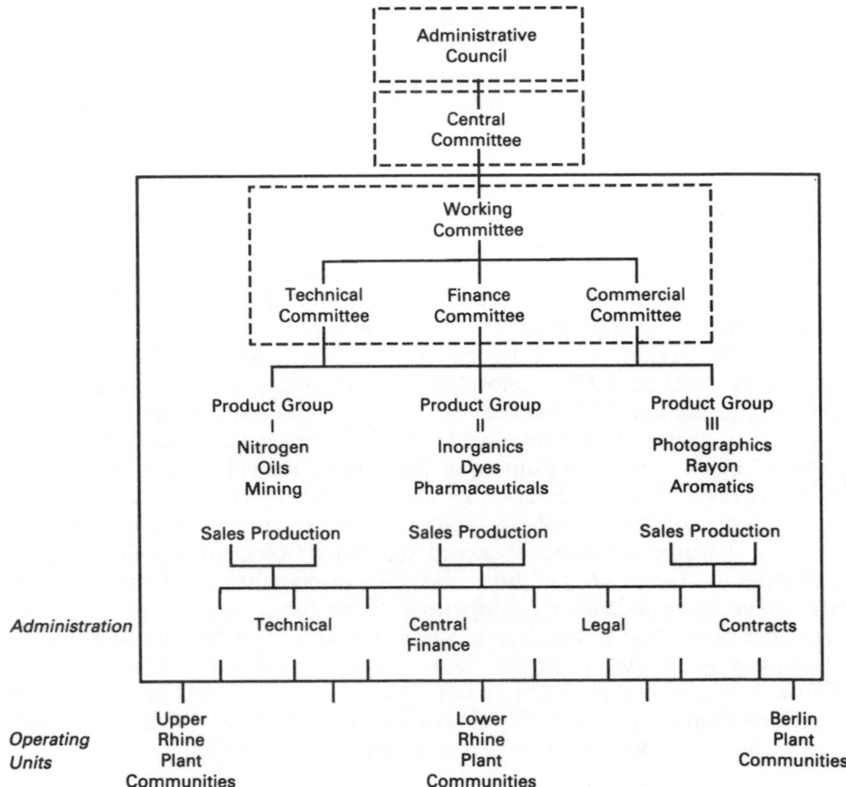

Figure 13.1 The organization of IG Farben, 1930

Source: Simplified and translated from Helmuth Tammen, *Die IG Farbenindustries Aktien-Gesellschaft, 1925–1933: Ein Chemiekonzern in der Weimarer Republik* (Berlin: H. Tammen, 1978).

Needless to say, industry diversification *per se* is a generic strategy of many enterprises in all types of modern industrial economies. In the case of technologically advanced multidivisional firms, however, the basis of diversification is their core technology, which they exploit in related industries (Chandler, 1977, 1990). An example of I.G. Farben, a diversified chemical giant of inter-war Germany, can be seen in Figure 13.1. Diversification becomes an offensive weapon whose industry sphere is somewhat focused. In the case of late industrialization, on the other hand, firms without a core and related technological capability diversify into technologically unrelated or remotely related areas. (See Figure 13.2 for the organization of the Nissan *zaibatsu*.) Diversification becomes a necessary but defensive tactic for growth. Thailand's large business groups provide a typical example of

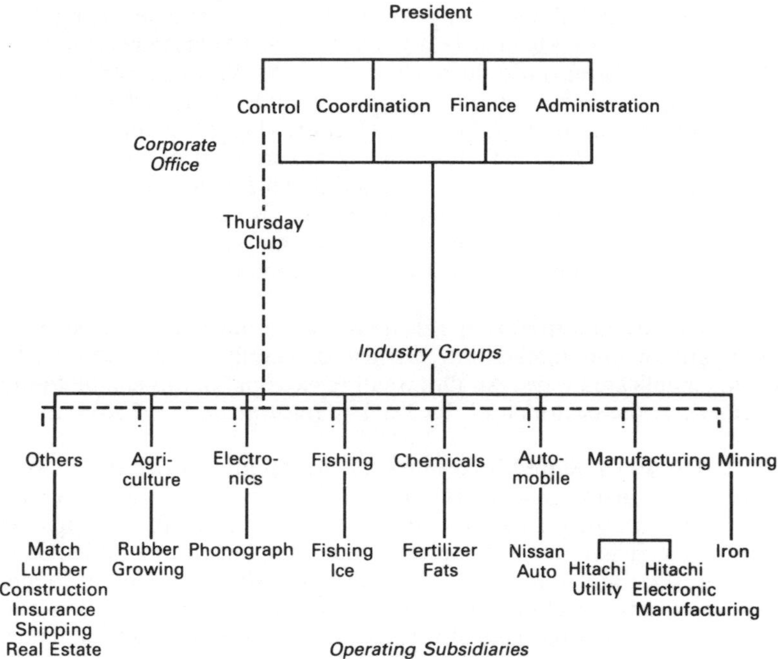

Figure 13.2 The organization of Nissan zaibatsu, 1937

Source: Based on the figures and descriptions of Masaru Udagawa, *Shinko Zaibatsu* (Tokyo: Nihon Keizai Shinbunsha, 1984), pp. 52–9.

unrelated diversification. In the 1980s the Yip In Tsoi group was involved in chemicals and automobile assembly. The SPI group manufactured textiles, detergents, frozen shrimps and store decorations. The industries covered by the Saha Union Group ranged from polyester textiles to machinery and spare parts, to construction services, to limousine services, to canned sea foods (Suehiro, 1985).

The trend towards diversification is further intensified by 'band-wagon effects' in oligopolistic behaviour: if one oligopoly diversifies into an industry which for the country in question is 'new', others feel compelled to follow suit to maintain parity in overall group size and strength (Miyazaki, 1980).[9] Because the 'new' technology is usually available from several international firms which constitute world technological oligopolies, several domestic firms also import variations of the same technological family and start competing against each other by improving them. Not only technology but oligopolistic structures are transferred worldwide. A classic case of this type of oligopolistic rivalry can be found in Japan's heavy electrical machinery sector. Two pioneers in the industry, both affiliated to the Mitsui *zaibatsu*, imported their technology from General Electric of the USA (the two merged to form Toshiba in 1939). Then in 1923 Mitsubishi established a heavy electrical business with technological ties to Westinghouse; and Furukawa, the copper mining and refining giant, founded Fuji Electric Manufacturing as a joint venture with Siemens (Uchida, 1980; Watanabe, 1984). In still later industrializers, the provision by government of subsidies to develop a new industry (new to the late industrializer in question, but most likely mid-tech by international standards) was also an incentive for oligopolistic band-wagon rivalry, as in the rush of the big three South Korean business groups to enter the motor vehicle, shipbuilding and heavy machinery industries.

In reality, late-industrializing enterprises have pursued strategies of vertical integration and unrelated diversification simultaneously, in unsystematic and complicated ways. An illuminating example of this can be found in the remark of the chairman of the Lucky-Goldstar group of Korea:

> My father and I started a cosmetic cream factory in the 1940s. At the time, no company could supply us with plastic caps of adequate quality for cream jars, so we had to start a plastic business. Plastic caps alone were not sufficient to run the plastic-molding plant, so we added combs, toothbrushes, and soap boxes. This plastics business also led us to manufacture electrical and electronic products and telecommunication equipment. The plastics business also took us into oil refining which needed a tanker-shipping company. The oil-refining company alone was paying an insurance premium amounting to more than half the total revenue of the then largest insurance company in Korea. Thus, an insurance company was started. *This natural step-by-step evolution through related businesses* resulted in the Lucky-Goldstar group as we see it today. (Harvard Business School, 1985; italics added)

Lucky-Goldstar's growth path illustrates the blurry division between vertical integration and technologically unrelated diversification. The point, however, is that in the mind of Lucky-Goldstar's chairman, his strategy was 'the natural step-by-step evolution through related businesses'. Each step, however carefully planned, was still in practice a leap technologically into another industry in which his company had no manufacturing know-how or capabilities. This diversification strategy had to be supported by incremental process innovation in each individual business, or the group would remain a mere collection of inefficient units. When, however, cumulative improvements were simultaneously generated in many industries, firms were able to exploit a certain degree of technological and organizational linkage through the sharing of usable knowledge (See Kagano *et al.*, 1985, ch. 4, for the Japanese cases.) If nothing else, they excel in the routinization of a strategy of diversification itself. As in the case of the Korean *chaebol*, they developed an investment capability to borrow the best foreign technology at the right price, minimize start-up delays and other costs, and ultimately improve whatever foreign technology they had borrowed.

Operational Characteristics

Top management behaves differently in the group structure just described compared with the managerial enterprises in industrial economies described by Chandler (1990). As stated earlier with respect to Carnegie Steel, and as implied in the organization of I.G. Farben (see Figure 13.1), top management, based on its technical as well as administrative capabilities, strategically controlled all divisional units (Hayes, 1987). Top management was not only concerned with the allocation of resources among the units, but was also involved in the manufacturing efficiency of unit operations. (See Chandler and Salsbury, 1971, and Hounshell and Smith, 1988, for detailed examples of Du Pont.)

In the case of the group structure of late industrialization, on the other hand, the function of top management is more or less confined to the allocation of resources among the units (product divisions or subsidiaries) and the monitoring of their performance (See Morikawa, 1976, for the Japanese cases.). Each business unit is responsible for the function of productive efficiency (see the example of the Nissan *zaibatsu* in Figure 13.2). This functional separation between top and operational management resulted from two factors. First, as long as the competitive weapon of late industrialization was incremental improvements on the shop floor, top management, by definition, could not make direct contributions. Second, given the nature and the speed of integration and diversification, top management could not acquire and exercise systematic knowledge of the firm's widely diversified products.

Unrelated diversification and functional separation between top and divisional management are also major characteristics of the conglomerate enterprises that have emerged mainly in the USA since the 1960s. In terms of technology, the conglomerates faced the same situation as late-industrializers. Both had no core technology with which to compete insofar as the conglomerates mostly originated in industries such as railway, mining and textiles whose technology was dead-end. These structural similarities notwithstanding, significant differences in terms of operations distinguish the two types of business enterprise.

Conglomerates tend to run their divisional units as independent entities with respect to finance, personnel and technology (Winslow, 1973). When the opportunity arises, therefore, top management does not hesitate to sell a unit to another firm. This type of transaction-oriented behaviour is infrequently seen among diversified business groups in late industrialization, which maintain closer linkages among business units. Even if individual operating units become relatively autonomous financially, top management still controls their major personnel decision-making. Middle and plant-level managers are regularly rotated among the units, through which top management encourages the transfer of administrative and technological knowledge. Thus troubled business units stand a better chance of being nursed back to health rather than sold off.

Foreign Operations

If late-industrializing firms adopted a strategy of diversification relatively *early* in their corporate histories compared to American or European modern industrial enterprises, they also were relatively *late* in multinationalizing their manufacturing operations.

A major strategy of industrial enterprises of advanced economies has been jumping overseas in the form of a manufacturing investment to exploit a technological or organizational competitive asset. Tariffs overseas played only a secondary role as the motive of direct investment (Kindleberger, 1969; Hymer, 1979). Thus, Singer Manufacturing Company, one of the earliest multinational enterprises, established its first overseas factory in Glasgow, Scotland, in 1867, only 16 years after Isaac Singer founded the company for commercializing his patented sewing machine, and in 1874 the company sold more than half of its products overseas (Wilkins, 1970, p. 43; Carstensen, 1984, p. 75; Hounshell, 1984, pp. 82–4; 93).

On the other hand, because individual late-industrializing firms do not possess a technological edge, they typically rely on exports to serve foreign markets and delay manufacturing overseas directly (at least in the markets intended to consume their outputs, rather than supply their inputs). This is especially the case in foreign markets in which technologically advanced domestic firms occupy the dominant position. Moreover,the competitive asset of late-industrializing enterprises, incremental process, and product improvements, is often too tacit to be transferred overseas easily, so only significant trade barriers provoke them to start overseas operations as a defensive measure. A telling example of this is Nissan Motor Company. This Japanese car maker, originally established in 1933 within the Nissan *zaibatsu* (see Figure 13.2), had various ties to foreign manufacturers, notably Austin Motors of Great Britain. Nissan started its overseas marketing activities early in its history, and by 1969 aggregate exports topped one million units. The company started its overseas manufacturing in Mexico in 1961, but for the production of its most significant export market (the USA), Nissan waited to invest until 1983, a 50-year lag after its founding, when trade restrictions became too heavy to bear. Nissan's competitor, Toyota Motor, an affiliate of the Mitsui group, followed a similar pattern. It began exporting cars to the USA in 1957, but only began operating in the USA (in a joint venture with General Motors) in 1984 (Udagawa, 1985; Toyota Motor, 1988). By contrast, the Ford Motor Company, originally incorporated in 1903, began overseas manufacturing first in Canada in 1904 and then in Britain in 1911. By 1914 the company's Model T had become the best selling car in the British market (Wilkins and Hill, 1964, pp. 16–18, 46–8, 51; Wilkins, 1970, p. 97).

The preference of late industrializers to serve foreign markets through exports rather than manufacturing investments resulted in the establishment of a unique organization: the *general trading company*. Since the late-industrializing firm's products were scattered across many industries, and because overseas trade (particularly exports), occupied a critical position in the firm's profitability, late-industrializing firms have formed subsidiaries to handle the trading activities of their entire group. For instance, the most prominent of the pre-war Japanese *zaibatsu*, Mitsui, founded a general trading company, Mitsui & Co., in 1876 as the group's import and export activities expanded after the Meiji Restoration. Within a few years it established overseas branches in Shanghai, Hong Kong, New York, London and Paris, handling a wide variety of agricultural, mining and manufacturing products. By the early twentieth century the trading company handled one-fifth of Japan's total trade (Mitsui & Co., 1977, pp. 23–54). Although Japan's trading companies became famous because of their size and scope of activities, general trading companies were also established in Malaysia, Thailand, the Philippines, Taiwan, South Korea, Turkey and Brazil (Junid, 1980; Cho, 1987).

CONCLUSION

Our major purpose has been to develop a paradigm of twentieth-century industrialization, at the core of which is borrowing technology that has already been commercialized by firms in more advanced countries. Whereas a driving force behind the First and Second Industrial Revolutions was the innovation of radically new products and processes, no major technological breakthrough has been associated with twentieth-century industrializers. The imperative to learn from others, and then realize lower costs, higher productivity and better quality in mid-tech industries by means of incremental improvements, has given otherwise diverse twentieth-century industrializers a common set of properties. Because the leading enterprises of these countries initially had no proprietary technology and could not compete in mid-tech industries against more experienced firms from advanced economies on the basis of low wages alone, their governments had to be more interventionist and developmental. In the most successful cases, government discipline of business was much greater than in the past. As process improvements constituted their major competitive strategy, manufacturing capabilities on the shop floor became their critical focus. Since individual enterprises did not possess any core proprietary technology, they grew by integrating and diversifying into unrelated industries. As a result, their firm structure featured business groups with technologically isolated divisional units.

From the viewpoint of a particular country, however, this industrialization process contains a paradox. The quicker a country learns and the closer it approaches the world technological frontier, the sooner it exhausts the opportunities to grow further by borrowing.[10] In the 'learning' paradigm the precise reason for success creates the very condition for dysfunction of this particular growth mechanism, whereas in the 'innovating' paradigm, the growth mechanism can, in theory, be sustained indefinitely.

Table 13.2 Balance of trade in mid-tech manufactured products of the USA and Japan, 1970–85 (millions of current dollars)

Year	*USA*	*Japan*
1970	2 911	4 305
1975	13 053	17 170
1980	6 772	52 184
1981	6 078	65 134
1982	–4 285	58 003
1983	–19 970	63 406
1984	–41 301	74 004
1985	–59 698	81 527

Note: The classification of mid-tech (medium R&D intensity) is described in OECD, *Selected Science and Technology Indicators: Recent Results, 1979–1986* (Paris: Organization for Economic Co-operation and Development, 1986).

Source: Adopted from UNCTAD, *Trade and Development Report, 1987* (New York: United Nations, 1987), Annex Table 8.

One can only speculate about how the learning paradigm will play itself out in the long run, as individual learners get closer and closer to the world technological frontier. Will they become innovators and, if so, will they become like other innovators or remain distinct?[11] Clearly as a country's economy matures, the effects of its early history on its behaviour weaken and new influences, both internal and external, begin to take hold. Nevertheless, to the extent that 'history matters', we may expect *some* early influences on a country's behaviour to persist, for better or worse.

Two congenital characteristics of twentieth-century late industrialization are a greater degree of state intervention than in earlier industrializations, and a greater propensity to regard incremental improvements in product and process as a competitive weapon. Japan provides the only source of evidence for the proposition that early technology history matters because it, along among late industrializers, has exhausted the opportunities to borrow technology in order to grow. Despite its status as innovator, however, its government continues to intervene more in promoting high-tech and monitoring firms than governments in other economically advanced countries, such as the USA. Trade balances in the 1980s in mid-tech industries show American manufacturing suffering and Japanese manufacturing earning the substantial surplus necessary to invest in high-tech (see Table 13.2). Evidently Japanese firms still emphasize incremental process improvements, notwithstanding their ambition and efforts to generate radical technological change.

Notes

Amsden presented another version of this chapter at the conference which inspired this book. She is grateful for helpful comments from Alfred Chandler, Paul David, Lily Hagen, Tom Hughes, Charles Sabel and John K. Smith. This version benefited from editorial help as well as substantive comments from Ross Thomson.

1. Gerschenkron did not systematically formulate his ideas on economic backwardness, but he suggested that because labour in backward countries is undisciplined and unproductive, the industries chosen by backward countries to promote must be those using the most advanced, labour-saving techniques (a disastrous policy in the 1960s from the viewpoint of employment creation). Backward countries then leap-frog to the technological frontier, leaving more industrialized countries behind. As for why these advanced countries do not bounce back, Gerschenkron writes that: 'either from inertia or from unwillingness to require or impose sacrifices implicit in a large investment program, [they hesitate] to carry out continual modernizations of their plant' (p. 10). (For an attempt to systematize Gerschenkron's ideas, see Roehl, 1976, and Sandberg, 1982).

   In spite of its fame and popularity, Gerschenkron's backwardness 'theory' has never been tested in a systematic manner, probably because it was not presented in a coherent form which allowed it to be tested empirically. Limited tests have generated mixed results, at best (Barsby, 1969; Good, 1973; Gregory, 1974).
2. Vernon (1966), in his product cycle analysis, implies that as an industry 'matures' it is more profitable for advanced countries, particularly the USA, to invest in creating new industries, and relinquish old ones to lower-wage countries (possibly through direct foreign investment). In this chapter we examine technology in terms of 'post-adolescent' rather than maturing, and industries in terms of 'mid-tech' rather than mature. The term maturation has the connotation of something being beyond its prime, but global demand in mid-tech industries may be growing and, by investing in incremental process improvements and production-related R&D as well as modern plant and equipment, these industries may substantially delay or even skip the stages of maturation and senescence which are logical consequences in the product cycle theory.
3. Pre-industrial artisan or craft skills, if transformed and integrated into modern economic systems, can be utilized as a significant asset for international competition. In Europe, particularly in France and Italy, industrialization benefited from the speciality products whose manufacturing was based on this type of skill, and whose unique market niche was eventually protected. To what degree such craft skills could and still can be successfully applied to modern industries, however, has been a matter of controversy. See Piore and Sabel (1984).
4. The government and the market share the role of disciplinarian, depending on the phase of an industry's development. During its early import substitution, the government's role tends to be most important. When the industry begins to export, the market mechanism substitutes for the disciplinarian. Then the government's role may increase as the industry restructures and ratchets up into a higher quality market niche, after which the market mechanism again takes command (Amsden, 1991a).
5. Technology transfer basically takes one of two forms, although there is eventually convergence. In industries where the necessary technological capability is within the borrowing firm's grasp, technology transfer typically takes the form of

imitating, or 'reverse engineering'. In many mid-tech industries, on the other hand, where the necessary technological capability tends to be beyond the borrowing firm's grasp, technology transfer typically takes the form of buying foreign technology (say, technical assistance, a foreign licence or a turn-key transfer: see Kim and Lee, 1987). In this case, however, what the buyer pays for is what Lall (1987) calls 'know-how' rather than 'know-why', the latter being an understanding of the fundamental technical principles underlying the technology. To gain that understanding will often involve the borrower (after buying foreign technology) in a process of reverse engineering to unlock the deeper structure of the technology. Another form of convergence in the two acquisition paths occurs as the firm which started by imitating often has to buy foreign technology to advance further.

It is noteworthy that technology transfer was no guarantee of narrowing a productivity gap even in the nineteenth century. As Pollard (1981) noted: 'Right up to 1850 and 1860, continental centres frequently failed to achieve British productivity and economy *even when using apparently similar equipment*' (p. 182; italics added).

6. Manufacturing capabilities have always been a significant part of the competitive strategy of all types of manufacturing firms, both innovators and borrowers (see Hayes and Wheelwright, 1984; Porter, 1985; Lazonick, 1990). Here we try to articulate the historically bounded nature of the significance of the shop floor.
7. Regarding the roadmap, by the 1960s there was evidence available on which industries had the greatest linkages and which had developed in what type of country (see, for example, Chenery, 1960; and Chenery and Taylor, 1968).
8. Some recent studies cast strong doubts on the innovative capabilities of large industrial enterprises (see, for example, Scherer, 1984, Chs 9 and 11). For technical reasons, however, these studies used such *inventive* outputs (in the Schumpeterian sense) as patents to proxy *innovative* activities. Here we are mainly concerned with the firm's ability to commercialize individual inventions, rather than the original sources of those inventions.
9. This phenomenon of oligopolistic behaviour was first examined by Yoshikazu Miyazaki as the 'one set-ism' or 'complete set principle' of the industry entry behaviour of post-war Japanese enterprises (Miyazaki, 1980). The same principle, however, can be found in the international entry pattern of world oligopoly members. See an example of the entry into the USA by the major European rayon producers in Markham (1952).
10. As far as individual industries are concerned, continuous borrowing naturally depends on the pace of new product and process innovations on the part of technology-generating firms in advanced countries. In some industries, such as chemicals and pharmaceuticals (in which world technological oligopolies are notably innovative), the conventional 'innovation first, borrowing second' pattern continues but, in others, the pattern has been changing rapidly. See Sato (1986) for Japan's present dilemma as a pioneering borrower who is now trying to transform itself into a major innovator.
11. A growing literature on convergence pioneered by Abramovitz (1986) suggests that growth rates and levels of productivity among advanced industrialized countries (including Japan) are becoming similar. Gomulka (1971) found a general pattern of a decline in countries' growth rates as they approached the world technological frontier. In this chapter we are concerned with the institutional arrangements that influence the growth mechanism in two sets of economies, which may or may not converge even as productivity levels equalize.

References

Abramovitz, Moses (1986) 'Catching Up, Forging Ahead, and Falling Behind', *Journal of Economic History*, XLVI, 2 (June).

Amsden, Alice H. (1977) 'The Division of Labor is Limited by the *Type* of Market: The Taiwanese Machine Tool Industry', *World Development*, March.

—— (1985) 'The Division of Labor is Limited by the *Rate of Growth* of the Market: The Taiwan Machine Tool Industry in the 1970s', *Cambridge Journal of Economics*, 9, 3 (September).

—— (1989) *Asia's Next Giant: South Korea and Late Industrialization* (New York and Oxford: Oxford University Press).

—— (1991a) 'Diffusion of Development: The Late-Industrializing Model and Greater East Asia', *American Economic Review*, May.

—— (1991b) 'Mind Over Matter: The Decline of the American Machine Tool Industry', mimeo, Department of Economics, Graduate Faculty, New School for Social Research, New York.

—— (1992) 'A Theory of Government Intervention in Late Industrialization', in L. Putterman and D. Rueschemeyer (eds), *The State and the Market in Development* (New York: Lynn Rienner).

Aylen, Jonathan (1985) 'European Steel – is there a Technology Gap?', *Steel Times*, 213, 11 (November).

Barsby, S.L. (1969) 'Economic Backwardness and the Characteristics of Development', *Journal of Economic History*, 29, 3 (September).

Becker, William H. (1982) *The Dynamics of Business–Government Relations: Industry and Exports, 1893–1921* (University of Chicago Press).

Brody, David (1960) *Steelworkers in America: The Nonunion Era* (Cambridge, Mass.: Harvard University Press).

Carstensen, F.W. (1984) *American Enterprise in Foreign Markets: Studies of Singer and International Harvester in Imperial Russia* (Chapel Hill, NC: University of North Carolina Press).

Chandler, Alfred D. Jr (1977) *The Visible Hand: The Managerial Revolution in American Business* (Cambridge, Mass.: Harvard University Press).

—— (1990) *Scale and Scope: The Dynamics of Industrial Capitalism* (Cambridge, Mass.: Belknap Press).

—— and Salsbury, Stephen (1971) *Pierre S. DuPont and the Making of the Modern Corporation* (New York: Harper & Row).

Chenery, H.B. (1960) 'Patterns of Industrial Growth', *American Economic Review*, 50 (September).

—— and Taylor, Lance (1968) 'Development Patterns Among Countries and Over Time', *Review of Economics and Statistics*, 50 (November).

Cho, Dong-Sung (1987) *The General Trading Company: Concept and Strategy* (Lexington, Mass.: Lexington Books).

Clark, Gregory (1987) 'Why isn't the Whole World Developed?: Lessons from the Cotton Mills', *Journal of Economic History*, 47 (March).

Clark, Kim B. and Fujimoto, Takahiro (1991) *Product Development Performance* (Cambridge Mass.: Harvard Business School Press).

Coats, A.W. (ed.) (1971) *The Classical Economists and Economic Policy* (London: Methuen).

Dahlman, Carl J. and Sananikone, Ousa (1990) 'Technology Strategy in Taiwan: Exploiting Foreign Linkages and Investing in Local Capability', mimeo, Washington, DC, the World Bank.

—— and Valadares Fonseca, Fernando (1987) 'From Technological Dependence to Technological Development: The Case of Usiminas Steeplant in Brazil', in Jorge M. Katz (ed.), *Technology Generation in Latin American Manufacturing Industries* (London: Macmillan).

Dertouzos, Michael, Lester, Richard K. and Solow, Robert M. (1989) *Made in America: Regaining the Competitive Edge*, Report of the MIT Commission on Industrial Productivity (Cambridge, Mass.: MIT Press).

Fishlow, Albert (1987) 'Alexander Gerschenkron', in J. Eatwell *et al.* (eds), *The New Palgrave Dictionary of Economics* (London: Macmillan).

Friedman, David (1988) *The Misunderstood Miracle: Industrial Development and Political Change in Japan* (Ithaca, NY: Cornell).

Gerschenkron, Alexander (1962) *Economic Backwardness in Historical Perspective* (Cambridge, Mass.: Harvard University Press).

Gomulka, Stanislaw (1971) *Inventive Activity, Diffusion, and the Stages of Economic Growth* (Aarhus, Denmark: Institute of Economics, Aarhus University).

Good, D.F. (1973) 'Backwardness and the Role of Banking in Nineteenth-Century European Industrialization', *Journal of Economic History*, 33, 4 (December).

Goodrich, Carter (ed.) (1967) *The Government and the Economy: 1783–1861* (Indianapolis: Bobbs-Merrill).

Goold, Michael and Campbell, Andrew (1987) *Strategies and Styles: The Role of the Centre in Managing Diversified Corporations* (Oxford: Basil Blackwell).

Gregory, Paul (1974) 'Some Empirical Comments on the theory of Relative Backwardness: The Russian Case', *Economic Development and Cultural Change*, 22, 4 (July).

Harvard Business School (1985) 'Goldstar Co., Ltd.', Case Study 9–385–264 (Boston, Mass.: Harvard Business School Case Services).

Hayes, Peter (1987) *Industry and Ideology: IG Farben in the Nazi Era* (Cambridge University Press).

Hayes, Robert and Wheelwright, Steven (1984) *Restoring Our Competitive Edge: Competing Through Manufacturing* (New York: Wiley).

Henderson, W.O. (1975) *The Rise of German Industrial Power, 1834–1914* (London: Temple Smith).

Hounshell, David A. (1984) *From the American System to Mass Production, 1800–1932* (Baltimore, Md.: Johns Hopkins University Press).

—— and Smith, John Kenly (1988) *Science and Corporate Strategy: DuPont R&D, 1902–1980* (New York: Cambridge University Press).

Hubbard, G.E. (1938) *Eastern Industrialization and its Effects on the West* (Oxford University Press for the Royal Institute of International Affairs).

Hughes, Thomas P. (1989) *American Genesis: A Century of Invention and Technological Enthusiasm, 1870–1970* (New York: Viking).

Hymer, Stephen Herbert (1979) *The Multinational Corporation: A Radical Approach* (New York: Cambridge University Press).

Johnson, Chalmers (1982) *MITI and the Japanese Miracle: The Growth of Industrial Policy, 1925–1975* (Stanford, Ca: Stanford University Press).

—— (1988) 'The Japanese Political Economy: A Crisis in Theory', *Ethics and International Affairs*, 2.

Junid, Saham (1980) *British Industrial Investment in Malaysia, 1963–1971* (Kuala Lumpur: Oxford University Press).

Kagano, Tadao, Nonaka, Ikujiro, Sakakibara, Kiyonori and Okumura, Akihiro (1985) *Strategic vs. Evolutionary Management: A U.S.–Japan Comparison of Strategy and Organization* (Amsterdam: North-Holland).

Kim, Linsu and Lee, Hosun (1987) 'Patterns of Technological Change in a Rapidly Developing Country: A Synthesis', *Technovation*, 6.

Kindleberger, Charles P. (1969) *American Business Abroad: Six Lectures on Direct Investment* (New Haven, Conn.: Yale University Press).

Lall, Sanjaya (1987) *Learning to Industrialize: The Acquisition of Technological Capability By India* (London: Macmillan).

Landes, David S. (1969) *The Unbound Prometheus, Technological Change and Industrial Development in Western Europe from 1750 to the Present* (New York: Cambridge University Press).

—— (1992) 'Homo Faber, Homo Sapiens: Knowledge, Technology, Growth, and Development', *Contention*, 1, 3.

Lary, Hal B. (1968) *Imports of Manufacturers From Less Developed Countries* (New York: National Bureau of Economic Research).

Lazonick, William (1990) *Competitive Advantage on the Shop Floor* (Cambridge, Mass.: Harvard University Press).

Maddison, Angus (1989) *The World Economy in the 20th Century* (Paris: OECD Development Centre).

March, Artemis (1989) 'The U.S. Machine Tool Industry and its Foreign Competitors', *The Working Papers of the MIT Commission on Industrial Productivity*, 2 vols (Cambridge, Mass.: MIT Press).

Markham, Jesse W. (1952) *Competition in the Rayon Industry* (Cambridge: Mass.: Harvard University Press).

McCraw, Thomas K. (ed.) (1986) *America versus Japan* (Cambridge, Mass.: Harvard Business School Press).

Mitsui & Co. (1977) *The 100 Year History of Mitsui & Co., Ltd.: 1876–1976* (Tokyo: Mitsui & Co.).

Miyamoto, Matao (1988) 'The Development of Business Associations in Prewar Japan', in Hiroaki Yamazaki and Matao Miyamoto (eds), *Trade Associations in Business History*, International Conference on Business History 14, Proceedings of the Fuji Conference (Tokyo: University of Tokyo Press).

Miyazaki, Yoshikazu (1980) 'Excessive Competition and the Formation of *Keiretsu*', in Kazuo Sato (ed.), *Industry and Business in Japan* (White Plains, NY: M.E. Sharpe; the article was originally published in 1965).

Morikawa, Hidemasa (1976) 'Management Structure and Control Devices for Diversified Zaibatsu Business', in Keiichiro Nakagawa (ed.), *Strategy and Structure of Big Business* (Tokyo: University of Tokyo Press).

Musson, A.E. (ed.) (1972) *Science, Technology, and Economic Growth in the Eighteenth Century* (London: Methuen).

Nelson, Richard R. (1987) 'Innovation and Economic Development: Theoretical Retrospect and Prospect', in Jorge M. Katz (ed.), *Technology Generation in Latin American Manufacturing Industries* (London: Macmillan).

Nye, John V. (1991) 'The Myth of Free Trade Britain and Fortress France: Tariffs and Trade in the Nineteenth Century', *Journal of Economic History*, 51, 1 (March).

OECD (1990) *Industrial Policy in OECD Countries: Annual Review, 1990* (Paris: OECD).

Okochi, Akio and Yasuoka, Shigeaki (eds) (1984) *Family Business in the Era of Industrial Growth*, Proceedings of the Fuji Conference, The International Conference on Business History, 10 (Tokyo: University of Tokyo Press).

Ozawa, Terutomo (1974) *Japan's Technological Challenge to the West, 1950–1974: Motivation and Accomplishment* (Cambridge, Mass.: MIT Press).

Piore, Michael J. and Sabel, Charles F. (1984) *The Second Industrial Divide* (New York: Basic Books).

Pollard, Sidney (1981) *Peaceful Conquest: The Industrialization of Europe, 1760–1970* (Oxford University Press).

Porter, Michael E. (1985) *Competitive Advantage: Creating and Sustaining Superior Performance* (New York: Free Press).

Rhee, Yung Whee, Ross-Larson, B. and Pursell, G. (1984) *Korea's Competitive Edge: Managing the Entry Into World Markets* (Baltimore, Md.: Johns Hopkins University Press).

Roehl, Richard (1976) 'French Industrialization: A Reconsideration', *Explorations in Economic History*, 13.

Rosenberg, Nathan (1982) *Inside the Black Box: Technology and Economics* (New York: Cambridge University Press).

Sandberg, Lars G. (1982) 'Ignorance, Poverty and Economic Backwardness in the Early Stages of European Industrialization: Variations on Alexander Gerschenkron's Grand Theme', *Journal of European Economic History*, 11, 3 (Winter).

Sato, Ryuzo (1986) 'Japan's Challenge to Technological Competition and its Limitations', in Thomas A. Pugel and Robert G. Hawkins (eds), *Fragile Interdependence: Economic Issues in U.S.–Japanese Trade and Investment* (Lexington, Mass.: Lexington Books).

Scherer, F.M. (1984) *Innovation and Growth* (Cambridge, Mass.: MIT Press).

Suehiro, Akira (1985) *Capital Accumulation and Development in Thailand* (Bangkok: Chulalongkorn University Social Research Institute).

Taylor, A.J. (1972) *Laissez-Faire and State Intervention in Britain* (London: Macmillan, for Economic History Society).

Toyota Motor Company (1988) *Toyota: A History of the First 50 Years* (Toyota City: Toyota Motor Corporation).

Uchida, Hoshimi (1980) 'Western Big Business and the Adoption of New Technology in Japan: The Electrical Equipment and Chemical Industries, 1890–1920', in Akiro Okochi and Hoshimi Uchida (eds), *Development and Diffusion of Technology* (Tokyo: University of Tokyo Press).

Udagawa, Masaru (1985) 'The Prewar Japanese Automobile Industry and American Manufacturers', *Japanese Yearbook on Business History: 1985, 2* (Tokyo: Japan Business History Institute).

Vernon, Raymond (1966) 'International Investment and International Trade in the Product Cycle', *Quarterly Journal of Economics*, 80 (May).

Wagoner, Harless D. (1968) *The U.S. Machine Tool Industry From 1900 to 1950* (Cambridge, Mass.: MIT Press).

Wall, Joseph Frazier (1970) *Andrew Carnegie* (New York: Oxford University Press).

Watanabe, Hisashi (1984) 'A History of the Process Leading to the Formation of Fuji Electric', *Japanese Yearbook on Business History*, 1 (Tokyo: Japan Business History Institute).

Wilkins, Mira (1970) *The Emergence of Multinational Enterprise: American Business Abroad from the Colonial Era to 1914* (Cambridge, Mass.: Harvard University Press).

—— and Hill, Frank E. (1964) *American Business Abroad: Ford on Six Continents* (Detroit: Wayne State University Press).

Winslow, John F. (1973) *Conglomerates Unlimited: The Failure of Regulation* (Bloomington, Ind.: Indiana University Press).

14 Epilogue: Institutions, Learning and Technological Change

Ross Thomson

However disparate in content and method, the contributions in this book have all interpreted technological change as a process of institutionally structured learning. Supportive institutions – firms, occupations, patenting systems, governments and universities – foster the acquisition of knowledge in ways that lead to changing techniques. Learning is thus the medium of ongoing productivity increase. Not all institutions support such learning; indeed, some obstruct it. The presence, type and outcome of technological change therefore vary with the structure and activities of institutions. It is thus fitting to conclude by considering the ways institutions, learning and technological change are connected.

THE FIRM

For the innovating firm, new techniques involve several complementary activities: inventing or acquiring inventions from others, developing techniques to practicality, discovering appropriate needs, identifying and utilizing marketing methods that can diffuse the technique or its product, arranging financing, and building or adapting production facilities. Each of these activities involves learning. Moreover, these types of learning are linked. Advances in one activity can be hampered by deficiencies in others, while fostering changes in yet others. When firms are organized to accomplish this learning, they can become and remain industrial leaders.

As Chandler shows (Chapter 3), these learning processes were bound up with the birth and growth of the modern, managerial firm. The managerial firm neither was nor is the only kind of firm to develop new techniques, but it proved particularly adept at organizing the development of techniques. Its marketing innovations widely diffused new techniques or the products made with them, thereby spreading the benefits – and the knowledge – of new techniques. Its mass production techniques lowered costs and augmented sales, making full use of the possibilities opened up by the new technique. In turn, through the spread of the technique, the managerial firm grew rapidly and positioned itself for further growth.

Firms must learn how to learn; experience does not necessarily lead to learning. Lazonick (Chapter 10) argues that there are two key requirements for learning in the managerial firm. First, the firm must hire skilled labourers and train them in its specific technology, marketing procedures and routines. The firm must also organize to make use of employees' skills. Developing new techniques involves interaction between individuals with the right kinds of specialized knowledge, so the firm must foster communication between production, marketing and research staffs.

Feedback from the commercialization of invention plays a central part in such communication. Invention, of course, precedes commercialization, but it also continues afterwards. As Smith argues (in Chapter 5), the linear view of technological change as a one-directional process from invention to commercialization ignores critical reverse information flows. During or after the original innovation, needs can be identified that will trigger changes in the product or process. Except when technical change targets clear problems in existing processes or products, innovators cannot avoid uncertainty about future market niches. This is particularly true for major inventions; their effects, as radio and cellophane illustrate, are not anticipated until the invention has first reached the market. The communication network of successful innovators thus extends outside the firm to its customers and suppliers.

When the firm has effectively organized communication flows between marketing, research and production staffs, it embarks on several learning paths. Learning by doing, which leads to incremental improvements in the production process, depends on both management and labour. If firms overly rely on the skills of managers or on workers' self-training, they may limit their prospects for such learning. Learning by using can create knowledge of the capability of equipment which can improve equipment utilization and design. Learning by selling is another crucial path to technical improvement. When the firm has clear communication channels with its customers, overcoming initial product deficiencies becomes easier and quicker. Customers and sales staff can also help identify potential targets for new products.

Each of these learning paths gives direction to inventive activity. Learning by doing overcomes deficiencies in the flow of production. Worker involvement is especially important for reducing waste of materials, smoothing production flow and adding to product quality, as Peter Lazes demonstrated for the case of Xerox in a paper presented to the conference. Learning by using can improve equipment utilization in a capital-saving manner, as Nathan Rosenberg showed for commercial aircraft (1982). Thomson notes in Chapter 6 that learning by selling developed shoe machines to practicality, further refined them and directed attention to new machines. The interactions around sale and servicing can also lead to marketing improvements, increased scale of output and thus direct invention towards new mass production techniques.

Technological innovation spurs two kinds of positive feedback that support further invention. First, by reducing costs or improving products, it raises profits and hence potentially increases investment in innovation. Second, the process of inventing augments the innovative potential of the organization. Successful innovative experience adds to the firm's learning capacity in each of its divisions. It also improves coordination among these divisions. As Chandler argues, firms that have developed new products often institutionalize their capacity to learn in a separate division oriented towards R&D. This new body strives to identify promising inventions outside the firm, while also developing inventions to practicality. Growing retained earnings and improved learning capability together direct inventive efforts to diversification throughout the firm's core technology.

Institutional structures also help us understand the limits to technological change. First, firms may not be well organized for the innovation–growth cycle. The contrast Nell describes in Chapter 12 between craft and mass production firms provides an important example. Mass production firms, using large-batch or continuous process techniques to produce standardized goods, are organized to develop new mass production techniques and spread them widely, but the craft firm is not. Craft firms employ a skilled, specialized production team, perhaps aided by machinery, to form products in small batches, often to the customer's specifications. Team members use their technical knowledge to innovate, but do so in ways the team can implement, thus maintaining the small-batch production system. In this way, craft routines direct and limit technological change. The firm's sensitivity to customer needs, by shortening production runs, undercuts mass production. Lacking advantages associated with scale, craft firms tend to remain small, so that growth occurs through the formation of new firms. Even if a craft firm wanted to move to mass production, it may lack the necessary financing, technological skills or marketing competence. Successful craft production may therefore lock firms out of mass production.

Second, learning within a core technology normally slows over time. In part this is due to changing consumption patterns over the product cycle. When market penetration is in full swing, increasing customer feedback fuels innovations to overcome design problems, extend the range of applications or increase the scale of production. Later, the market grows more slowly while dominant designs and routines lock the firm into a given system. When markets and profits level off, these limits are reinforced.

Diversification may overcome some of these problems by creating new products that extend the range of the core technology. Then customer feedback again increases, profits rise, and the new learning process can again lead to technical improvements. But the same learning limits arise as the firm exhausts the possibilities of its core technology. Both Harris (Chapter 7) and Chiaromonte, Dosi and Orsenigo (Chapter 8) model this limit by assuming that, within technological regimes, dynamic increasing returns occur at a diminishing rate. Radical new technologies can overcome these limits, but firms usually have few advantages and much trepidation about innovating in such uncharted territory.

Here is a third limit to technological learning. Firms are far better at ongoing development than they are at basic invention, even within their core technology. A mass production firm is organized for large-scale production, distribution and incremental product changes, but has fewer specialized skills in undertaking basic invention. Agents outside the firm are responsible for most basic inventions. These may be new or small firms or, especially when more fundamental research is involved, universities and governments.

The dependence of firms' innovation on interaction with customers, on workers' training outside the firm, on the invention of other firms and on university and government research all exemplify the interdependence of institutions in technological change. One firm by itself cannot produce all the design, production, and marketing knowledge it needs to innovate. Fortunately, it does not have to.

INTERFIRM LEARNING

When some firms learn from the innovation of others, technological change is effectively cooperative. Nelson's distinction between generic and specific knowledge helps account for the extent and limits of this cooperation (Chapter 2). Generic technological knowledge is generally applicable and may be widely diffused in a supportive institutional setting. A positive externality for the economy, new generic knowledge may lead to an unintended cooperation among firms.

Specific technological knowledge, on the other hand, applies to a particular product or process and involves tacit knowledge embedded in the routines of the firm. Specific knowledge does not exclude interfirm learning. Indeed, intense communication between innovators and their customers is often central to the development of specific knowledge. Firms may also agree to share specific knowledge. However, because this knowledge is hard to separate from the personnel and practices of the organizations that generated it, it does not easily flow to competitors in the absence of interfirm agreements.

The specific character of knowledge allows firms to rely on lead time and complementary investments – rather than patents or secrecy – to protect their investments long enough to reap the benefits. Because an important part of firms' innovative activity is thus protected, their drive to invent is not inhibited by the diffusion of generic knowledge.

Interfirm learning is a basic aspect of technological change in capitalist societies. Even though firms develop their own innovative capabilities, they begin by appropriating generic knowledge held by the managers and labourers they hire. Learning between firms thus supports planning within firms. The most successful firms may well be those that quickly acquire and develop publicly available knowledge. They learn not only of generic improvements but also of outside patents and small firms that might be worth acquiring and expanding. They also learn of institutions, including the ways other firms organize marketing, production, and R & D. Finally, interfirm learning may lead to complementary innovations needed for the success of earlier inventions. The market for cellophane, for example, depended on the birth of the supermarket, and radio's success depended on the inception of broadcasting.

The magnitude of interfirm learning varies with the way firms organize technological change, as Thomson argues. When firms invent for their own usage, other companies learn of the inventions through altered products, mobility of labour or new firm formation. But when firms develop techniques for sale, their interest is to sell as widely as possible, which also spreads knowledge across the market. Greater diffusion fosters broader ongoing invention than does self-usage, for the knowledge communicated to potential inventors and other firms supports their invention. Wider diffusion also broadens the information flows back to the capital goods firm, giving direction to its ongoing invention. Profits from wider diffusion can also finance ongoing invention.

The importance of capital goods for ongoing invention is manifested in the behaviour of their inventors. Learning from the communications networks in which they are embedded, inventors for capital goods firms use their knowledge to invent more for the same and other firms. They not only invent more for the particular capital-good industry, but – making use of their generic knowledge – they also invent more in other industries. They thus become an important means of technological convergence.

The capital-goods industries unite technological change across whole economies. The machinery industry in the nineteenth century did this by carrying mechanization – and thus the realization of economies of scale – to many industries. Capital goods firms played this critical role by spreading generic technological principles and widely applicable products among many users. Because these firms, like other media of interfirm learning, operated more readily within regional networks, the rise of mass production in certain regions began to differentiate the economic development process among regions. Yet capital goods industries can also equalize levels of development among regions and nations. In certain circumstances, multinational corporations may have this effect. And, along the different path Amsden and Hikino call late industrialization (see Chapter 13), less developed countries can industrialize by purchasing technology from technological leaders.

The cooperative character of learning is an important feature of models of technological change. Semmler (Chapter 9) shows that learning among competing firms can limit the tendency to universalize a single technique, thus maintaining the diversity of techniques among firms. Chiaromonte, Dosi and Orsenigo develop a model involving both intra-industry learning and interaction between capital-goods producers and users. Their simulation results depend on the combination of both types of interfirm learning. Harris demonstrates that the dynamics of a sector which concentrates knowledge (such as the capital-goods industries of modern economies) can influence the rate of technological change within an economy and its convergence between economies.

Interfirm learning thus takes a variety of forms: learning between producers and users of a new technique, between potential producers, between producers of complementary goods in complex technological systems (such as cars, petroleum and heavy machine tools), and between producers and firms in relatively unrelated industries. Each type of learning involves feedback loops within and between firms that normally reinforce the conditions for technical change. Technical change also typically boosts firm profits while stimulating sectoral and aggregate demand, thus supporting further technical change.

However, in some cases learning may also have modest or even negative effects on technological change. Technical change may require a certain scale of economic activity. Harris (Chapter 7) and Parker (Chapter 11) both suggest that the capital goods industries need to acquire a certain size before they widely spread generic knowledge. Moreover, the structure of knowledge may direct attention away from fruitful solutions to technological problems; the whole economy may be locked into concepts and routines that limit productivity. This is particularly clear for industry standards, such as the heavy automobile, the English measurement system and the QWERTY keyboard. Interfirm learning among craft firms may reinforce their commitment to craft techniques, thus directing them away from mass production. Finally, interfirm learning that is too quick and inexpensive can undercut the rationale for invention.

LEARNING AND INTERFIRM COMPETITION

Due to the firm-specificity and uncertainty of learning, technological change is a source of variation among firms. Differences in technique are therefore normal, resulting from the distinctiveness of firm histories. Furthermore, distinct learning capabilities mean that firms face different costs in developing techniques, different prospects for success and thus different inclinations to engage in innovation. Those firms with large, well-structured investments in plant, marketing and industrial research face the fewest hurdles in developing new techniques, especially when related to their core technologies.

Greater learning capabilities can thus convey competitive advantages that can lead to increased market shares. Chandler maintains that superior capabilities in mass production and scientific industries conferred advantages on German and American firms over their British counterparts, and these advantages were used to increase their share of world markets, including the domestic British market. Similarly, Lazonick notes that Japan's advantages in labour training may lead its firms to win market shares from US competitors.

However, learning may also work against competitive differentiation. Firms imitating the innovations of others can reduce or even eliminate their competitive disadvantages. The firm-specific nature of technological knowledge limits such imitation: it will be a costly, time-consuming, and uncertain endeavour. The advantages of early movers are reinforced by their investments in production and marketing, which allow firms to forestall competitors' entry by lowering prices, improving quality or servicing, or further changing technology. Nevertheless, if the full range of imitation costs are lower than invention costs, the path of diffusion can eventually lead to more uniformity of technique and competitiveness.

Learning, then, affects the competitive struggle. Competitive structure also affects learning and hence technical change. The costliness of technological change may constrain innovation in industries with many small firms. Even large firms may be put at a temporary competitive disadvantage by high innovation costs. Because high innovation costs may persist for long periods and may never be recouped, this disadvantage can undercut the innovation. Inventing firms therefore need some protection, be it market limitations (perhaps by tariffs) or abundant financing (coming from credit or profits in other sectors).

The introduction of endogenous innovation and diffusion adds substantial complexity to models of competition and technological change. Economists often resort to simulations to simplify the task. Such models, constructed by Semmler and by Chiaromonte, Dosi and Orsenigo, demonstrate that subtle changes in competitive structure, normal practices of firms and the cost and ease of invention and diffusion can strongly affect productivity growth and market shares. In some cases, positive feedback can reinforce initial technical changes, leading to steady productivity growth. In others, technological change is hampered and even extinguished. Simulation results do show that regimes of price-taking by firms generally have lower rates of technological change.

Competitive structure can dictate the form that successful learning must take. As Amsden and Hikino argue, late-industrializing countries, without the institutional capability to develop new techniques, must transfer technology and organizational models from abroad. But they do not yet have organizations easily able to build firm-specific knowledge and routines around transferred technology. Late industrializers must therefore establish distinctive firm (and group) organizations to create learning capabilities at the same time as they undertake incremental shop-floor improvements. Normally governments of late industrializers also direct and insulate these key firms through controls, procurement policies and subsidies. These activities may have one unintended effect: the institutions developed for catching up may provide a superior basis for innovation later on. The form of competition thus evolves.

TECHNOLOGICAL COMMUNITIES

Technological change also depends on learning outside the myriad forms taken within and between firms. Other institutions organize people into what Nelson calls technological communities. Professional associations, universities, governmental bodies and occupations can all serve as technological communities. These communities communicate technological knowledge among their members. Since members are often associated with firms, technological communities are media of interfirm learning. Members also use new-found knowledge to invent. Technological communities are thus important for realizing the positive externalities that arise from convergent technologies.

Technological communities often bridge economic and scientific institutions. As Parker suggests, science, technology and the economy develop in interconnected but separate streams. Each has its own institutions, practitioners and channels of communication. Technology has advanced through a series of communities that are increasingly connected to science. Whereas craftsmen with relatively little scientific knowledge propelled the Industrial Revolution, modern technological change requires a distinctive profession, engineers, who mediate between scientific advance and economic need.

The patent management system described by Cooper in Chapter 4 was one important technological community. It organized key activities for inventors: securing, marketing, enforcing and renewing patents. Combining economic, political and intellectual elements, this system functioned as a nationwide 'invisible college', communicating knowledge about inventions and inventing and thus developing technology. Social interactions surrounding the patent system tied diverse groups together, including patent examiners, patent agents, patentees, assignees, mechanical experts, judges, juries and often the press. Journals spread knowledge of patents across an even wider domain. Selling and licensing patent rights soon became a business in itself, further extending the reach of this community.

The patent system not only spread knowledge of particular inventions, but it also helped to define what an invention was. Discussion and debate to identify the novel aspect of a new technique helped to form a common set of categories describing technologies and how they can change. This understanding began to shape technological knowledge into a system of ideas shared by community members. As a result, common routines for inventing were encouraged, while the identification of convergent technologies became easier. Whatever its effects on incentives to invent, the patent system spread technological knowledge widely, and was thus an important institution of learning. This learning, and the revenues coming from patent sales, fostered ongoing technological change.

Governments shaped not only patenting but a number of other technological communities. They were particularly important for developing technologies for which externalities would have made private development difficult. From their decisive role in organizing interchangeable-parts firearms production in the early nineteenth century, to their support for computer and aircraft development in the Second World War, to the Department of Defense's recent wide-ranging efforts, US military needs have shaped highly important technologies. Civilian government offices as different as US agricultural experiment stations, Japan's MITI and South Korea's state-owned steel company all developed and diffused technologies. The grant-reading and writing community has contributed substantially to the development of basic sciences, applied sciences and medicine. Furthermore, public education has organized technological communication in a variety of ways, including the development of technological professions.

University-trained engineers formed a community that played a powerful role beginning in the late nineteenth century. Engineers carried the generic knowledge of applied science (scientific understanding and procedures as well as empirical knowledge associated with their engineering experience). In the twentieth century, the conceptual system in which engineers were trained has become the main intellectual context in which technological problems are posed and solved. The evolution of engineering education is therefore an important determinant of technological change. Growing hand-in-hand with corporate industrial research laboratories, the engineering community affected the form, content and success of companies' technological change.

Sometimes interaction within technological communities generated new technologies, standards and routines. For example, the meetings of the Franklin Institute in Philadelphia helped to establish the US standard screw thread in the 1850s. Governments, too, have set industrial standards. Communication around patent agencies or engineering schools often directed invention. Such direction was particularly important for radical inventions. In a talk in the conference (see also Hughes, 1989), Thomas Hughes suggested that, unlike the incremental inventions occurring in large technological systems, fundamental new inventions were characteristically undertaken by independent inventors. These inventors were products and parts of various technological communities.

Finally, community members, making use of their networks, often founded firms to develop and sell new products. Following the practice and using the knowledge of their communities, patentees, machinists and engineers formed firms that began the process of development and diffusion. More often than not, the pioneers failed, but their efforts were necessary for others – including the corporations Chandler describes – to succeed.

Technological communities thus played several significant roles in the process of technological change. Most evidently, they communicated technological knowledge, both of particular techniques and of general categories. They also taught people about the process of inventing, patenting and gaining use for inventions. They generated and spread some technologies, and many of their members invented and formed firms to gain the benefits of their inventions. Technological communities developed in close contact with private firms, and were major contributors to technological change within these firms. In some cases, they were even more important; without chemists and chemical engineers, there would be no mass production chemical industry.

Of course, such communication did not make knowledge a universally available and costless public good, for knowledge diffused most readily within technological communities and not as easily outside them. We should not be surprised, therefore, to find that technological change remained concentrated within the occupational and geographic limits of these communities. As such, new technological communities are important components of the growth of leading industrial countries.

HISTORICAL EVOLUTION OF TECHNOLOGICAL CHANGE

Technological change is inherently historical. Dominant techniques and the institutions surrounding them succeed not because they are optimal – many are inferior to known alternatives – but because they were viable in the past and developed in cumulative learning processes that added to their advantages. Ongoing learning and increasing returns may lock in techniques or institutions even if superior possibilities are known. Thus, yesterday's institutions and technologies limit and direct today's technological changes. But institutional change can alter these limits and directions. To understand technological change, therefore, we must take account of the interactive history of institutions and technologies.

The chapters of this book take some important first steps. They identify a series of institutional changes that altered the way learning occurred and techniques changed. The transformation from the family firm to the emergent managerial enterprise to the diversifying managerial firm lies at the core of the process.

Neither the managerial firm nor its mass production came easily. The change involved, as Nell argues, major transformations in technology and in firm structure. The skilled work team of the family firm did not know mass production and mass-marketing techniques, and neither was it commonly interested in techniques that would undercut its own organization of work and family control over capital. Therefore, learning tended to continue along custom or small-batch technological trajectories.

Learning by the family firm was fundamental for economic development. The great technological changes of the Industrial Revolution were largely undertaken by family firms. By learning how to employ and manage dozens or even hundreds of workers within single plants, these firms were important new institutions in their own right. Differing fundamentally from the merchant-based putting-out system, they were a first step on the road to mass production. But learning within the family firm did not achieve the scale and scope economies of modern capitalism. In places, a British-style 'personal capitalism' resulted, skilled in developing custom products but deficient in the mass production strategies of the managerial firm. It is provocative to note that the type of firm can periodize the process of technological change.

The family firm operated in a context of other institutions, and the structure and evolution of these institutions in part explain the success and the decline of the family firm. Capital-goods firms were essential parts of this context. Together with markets for skilled labour, they gave rise to the external economies that, as Alfred Marshall noted, helped to form specialized industrial regions.

The capital-goods sector coming out of the Industrial Revolution transformed the process of technological change. Indeed, the emergence of a separate machinery industry is a useful way to divide the Industrial Revolution from what followed. This sector speeded up technological learning within industries and industrialized areas, then spread this knowledge to new areas and industries. Machinists and others in capital-goods firms commonly pioneered new techniques and the companies that commercialized them. Moreover, capital-goods firms invented many mass production techniques and diffused them within and between sectors; the nineteenth-century US companies that invented mass-production machine tools and sold them to many industries were a particularly important example. Capital-goods firms were also critical in internationalizing mass production in the twentieth century.

Personal capitalism had distinctive technological communities which fostered technical change. The craftsman-mechanic whose mobility diffused much of the technology of the Industrial Revolution persists to this day. But, during the nineteenth century, these mechanics were increasingly likely to found capital-goods firms rather than to use inventions in their own shops. The patent-management system played an especially important role in learning processes prior to the existence of a large machine-tool sector, R&D laboratories and university-trained engineers. This system widely disseminated technological knowledge. It also helped to shape a language of invention that did not require craft-specific knowledge, which thus made it easier to identify generic features of inventions. Patent agencies regularly communicated with capital-goods firms, and both invented widely. Inventors for these companies took out more patents, spread them across more industries and gained usage more readily than did other inventors. Their inventions included many mass production techniques.

Government activity was also significant for the difficult transition to mass production. Federal government purchases of, and support for, firearms made with interchangeable parts were critical factors in the rise of mass production in the USA. From about 1870, governments organized scientific education in German polytechnics and US land-grant colleges, which trained workers for the flourishing engineering trades.

Thus institutions of personal capitalism created conditions for the birth of a new, managerial, form of capitalism. Patent-management systems, labour markets, capital-goods firms and governments each contributed to the mass production and scientific technologies of the late nineteenth century. The very success of personal capitalism created other conditions by accelerating the rate of capital accumulation and hence market growth, and by integrating markets through transportation, communication and merchandising improvements. Infrastructural investments also helped form managerial prototypes and long-run capital markets that industry then utilized.

In this setting, industrial firms brought mass production technologies to fruition. New products and production processes gave these firms competitive advantages that supported their growth. Complementary marketing, production and managerial innovations realized scale economies that furthered firm expansion. Innovations in some sectors created technological knowledge which capital-goods firms, skilled workers and technological communities diffused to other sectors. Spreading mass production resulted also from the diversification of firms discovering and utilizing scope economies.

The difficulty of introducing the new technology and organization of managerial capitalism is shown by the firms that failed to negotiate the transition. Many firms were locked into limiting organizational forms. Chandler and Lazonick suggest that this was the case for British firms. Tied to its personal, craft structures, the British firm failed to integrate into marketing, to develop mass production, to organize a professional management or to train workers. New leaders, particularly US firms, gave birth to the new economic order.

Then the combination of diversifying corporations, research laboratories and engineering schools introduced new patterns of diffusion and technological learning. Institutions in the twentieth century continued to depend on market-mediated learning. Firms still used marketing knowledge to define their products, while capital-goods firms – now including firms designing whole plants – still developed mass production techniques. But more and more of the process of technological change relied on learning inside the firm's industrial research laboratories.

Innovative managerial firms could succeed because they operated in a rich and supportive institutional environment. Firms learned from each other. Private industrial research grew alongside the training of engineers. Pioneering companies provided important components of major new technologies. Europeans pioneered the car, the great fruits of which were until recently gathered by the USA. Moreover, unintended consequences from long-established trajectories were often decisive. The significant effects of electrification in the 1920s were preceded by a century of electrical development through the telegraph, light bulb, dynamo and utilities; and, as Paul David suggested (1989), the computer may resemble the dynamo in this regard. Governments undertook their own research, subsidized private research and stabilized the economy. The mass production firm was one among many novel institutions in a distinctive economic order.

The cumulative learning processes of this new order allowed the US economy to maintain and even expand its productivity advantages over the first half of the twentieth century. It and other Western economies have continued to expand these advantages in relation to most countries of the world. Their success has internationalized key parts of technological learning, which spurred productivity growth among countries that could absorb it. Education, scientific and engineering publication, technology transfer and transnational corporations have each helped latecomers surmount competitive disadvantages.

Viewed in its historical context, it is clear that the evolution that led to today's economy is no more finished than it was a century ago when the managerial firm was emerging. Like a century ago, successful latecomers have built new institutions to challenge the world's industrial leaders. Whether these institutions can lead technology in new directions will depend on the way they structure the process of learning.

References

David, Paul (1989) ''Computer and Dynamo: The Modern Productivity Paradox in a Not-too-distant Mirror' (Stanford: Center for Economic Policy Research Publication 172).

Hughes, Thomas (1989) *American Genesis: A Century of Invention and Technological Enthusiasm, 1870–1970* (New York: Viking).

Rosenberg, Nathan (1982) 'Learning by Using', *Inside the Black Box: Technology and Economics* (Cambridge University Press).

Index